E Lost Key

How we Overlooked the Best Idea of the 20th Century

Alexander Unzicker

Copyright © 2015, 2022 Alexander Unzicker
All rights reserved.
ISBN-13:
978-1519473431
ISBN-10:
1519473435

Contents

Introduction   7

Chapter 1   11
The Blind Spot of Physics
Why Einstein's best idea remains unknown to this day

Chapter 2   29
"I Rely on Intuition"
Why all scientific revolutions began without calculations

Chapter 3   41
"Arbitrary Numbers ... Ought Not to Exist."
Einstein on simplicity in nature

Chapter 4   55
A Precursor of Deep Thoughts
Relativity, Mach and the secret of the universe

Chapter 5   71
The Revolutionary Idea of 1911
Why light propagates along curved paths

Chapter 6 — 89
Seduced by Mathematical Beauty
The Pyrrhic victory of geometrical formulation

Chapter 7 — 113
Gravity From the Universe
Einstein could not see that Mach was correct

Chapter 8 — 129
Half a Century Too Late
Robert Dicke's simple completion of Einstein's idea

Chapter 9 — 149
The Genius Who Didn't Talk to Einstein
How Dirac's Large Number Hypothesis enters the game

Chapter 10 — 165
Big Bang Without Expansion
Einstein's desired cosmos

Chapter 11 — 179
Back Before Newton?
Why we need to question the notions of space and time

Chapter 12   **195**
Forgotten But Not Hidden
Einstein's idea is visible long since:
It is dark energy

Chapter 13   **211**
**The Next Revolution Needs
Open Data**
Why astronomical precision measurements
must go online

Outlook   **228**

Acknowledgment   **228**

Images   **228**

Bibliography   **229**

Index   **231**

Endnotes   **232**

Introduction

This book is about cosmology. You may ask whether we can afford to deal with abstract intellectual insights in view of the possibly greatest challenges humankind has to face in history. Will we be able to protect the Earth from cosmic catastrophes? Will we be able to preserve this wonderful planet from the severe consequences of the burdens we are placing upon it through our own negligence? Will changing societies succeed in upholding the basic principles that ultimately enabled human coexistence?

Will the evolution of our possibly unique intelligence that we manipulate substantially, modify our existence as human beings? Will there be other civilizations, if any, that follow? As history has shown, it has always been a profound understanding of reality that has constituted the first step of advancements made by civilizations. I am convinced that our long-term survival critically depends on how deeply we are able to reveal the laws of nature.

The incentive that has led to the greatest insights of humanity has never been utility but a commitment to understanding the world. One figure who was particularly passionate about discovering the secrets of nature was Albert Einstein, who one century ago contributed in an outstanding way to the progress of science.

It is commonly known that Einstein was a genius, though an incomprehensible one. Yet this was not *his* view. He rather insisted that if a person understood something thoroughly, he or she should be able to present it in plain words. I agree with him in this respect, and contrary to other popular science authors, I don't believe that you, the reader, is too stupid to comprehend Einstein's theories of relativity. Rather, I consider it the duty of a scientist to share this knowledge with those who have not had the privilege to dedicate their lives to the big questions.

This transfer of knowledge is not only an obligation of research, but is what I believe to be both a necessary and fruitful grounding for science itself. The history of science has shown many examples where groups of so-called experts have gone astray in their convictions, stalling progress for extended periods of time. Therefore, it is essential that our knowledge can be evaluated by people with general erudition in order to preserve its correctness for the long term. Society must understand and judge the theories that will decide its destiny.

There are indeed some hints that the currently dominant models of our understanding of nature are too contrived to be credible. Certainly, the stories about parallel universes and wormholes seem more a part of the science fiction industry than Einstein's physics. To which degree we are still on the right track is hard to say, and many would think that my personal views about this are too critical. But sheer logic should tell us that it makes sense to consider at least the possibility that we have been going astray for some time.

This is why I consider it worthwhile to excavate old ideas and work on them. It is not only an intriguing theory of Einstein's that I think needs to be reevaluated, but also other profound ideas that are closely related to it. The Viennese philosopher Ernst Mach, the Nobel laureates Erwin Schrödinger and Paul Dirac, the visionary cosmologist Dennis Sciama, and the American astrophysicist Robert Dicke, all had something to say, which in combination could revolutionize our current understanding of the universe. In view of all the evidence we have today, I am tempted to claim this all fits together like a beautiful puzzle. I am curious as to how you will value the arguments I provide. You may choose to agree or disagree. As a matter of fact, half a dozen of the great physicists could not link their thoughts as we can today.

It is unlikely that the message of this book will be received with open arms in the astrophysical community. I think this is as unintentional as it might be unavoidable. When a book implicitly attacks a canon that has been established for decades, one should expect a reaction, and thus I shall be prepared for critique of any sort.

However, it is fair to say that this synthesis would have attracted the attention of Mach, Einstein, Dirac, Schrödinger, Sciama, and Dicke. If nothing else, I am happy with that thought. And if there is no other merit of this book, for the first time it exposes that there is an unrecognized overlap of ideas that these physicists never talked to each other about. I hope I can excite you through exposure to this fascinating interplay and wish you a pleasant read.

Munich, November 2015

Alexander Unzicker

Some historically interesting facts that I was not aware of were included in the reprint. Recent observations of gravitational waves, dark energy and black holes were also briefly discussed.

Munich, July 2022

Alexander Unzicker

Chapter 1

The Blind Spot of Physics

Why Einstein's best idea remains unknown to this day

Einstein – we call him the greatest physicist of all time, the genius, the hero of science. He scoffed at the hype that surrounded him, yet as a scientist, Einstein was a one-off. Not only is a great deal of theoretical physics based on his discoveries, but Einstein's unique way of working, his uncompromising quest for the truth, and his implacable drive to understand nature, all make him a role model to this day.

So one might think that his achievements must have been studied down to the finest detail and his ideas are followed up by researchers throughout the world. Unfortunately, this is not the case. There are plenty of biographies, certainly, illuminating every hidden corner of his life - but the appraisal of his scientific legacy focuses on very few works that underlie his famous theories.

Particularly prominent are his theories of special and general relativity, which he formulated in 1905 and 1915, respectively. The picture of their origin however remains complete, if we blank out those ten years during which Einstein grappled with the theory almost continuously. History often reduces a complex storyline to what ultimately emerged. Sometimes this or that decisive factor is left by the wayside – and that is what seems to have happened with the origins of the theory of relativity: one simple yet revolutionary idea of Einstein's has been completely forgotten.

Einstein's Lost Key

Countless creative contributions from Einstein and his comrades-in-arms, featuring plenty of wrong turnings, gradually shaped the General Theory of Relativity. Seeing the version that emerged in November 1915 as a single discovery is not justified if we analyze its physical content in depth. Rather, it appears that around 1911, Einstein had in his hand the key to an even greater discovery, a ground-breaking idea that would have explained gravitation directly from the characteristics of the universe: a theory based on a variable speed of light. Not only would c, the speed of light, be affected by all the mass in the universe, so would the very definitions of the meter and the second. These then variable yardsticks of length and time would join to create the illusion that light travels at a constant velocity of 299,792,458 meters per second.

Having derived his special theory of relativity in 1905 from this measurably constant speed of light, Einstein generalized this theory in a stroke of genius. This revolutionary idea puts much of the physics of the 20th Century in doubt, up to what is called the accelerated expansion of the universe - and indeed the expansion itself. Initially, however, the consequences of Einstein's idea of variable time scales subtly conceal themselves. In 1911, he put it in a nutshell: [1]

> *"Nothing forces us to assume that... clocks have to be seen as running at the same speed."*

This article, published in the prestigious *Annalen der Physik*[j], has remained amazingly unknown over the years. Not only did this approach remain one blind spot in the field, people wrote about variable speed of light without even mentioning Einstein as the forefather of the idea! Those later attempts however overlooked the core of Einstein's discovery, which would have allowed the calculation of a fundamental constant of nature: New-

---

[j] Not to be confounded with the journal *Annals of Physics*.

Chapter1: The blind spot of physics

ton's gravitational constant would have been rendered superfluous. Einstein would not only have improved Newton's theory, he would have replaced it - and outdone his predecessor in an even more fundamental way.

A FLASH OF INSPIRATION BASED ON PURE THOUGHT

In fact, a variable speed of light was the initial stimulus that set Einstein off in the quest for the theory of general relativity. He mentioned it as early as 1907. In the same year he had formulated the principle of equivalence, a brilliant thought experiment that is recognized as the foundation of the theory. Einstein realized that it was impossible to distinguish between being driven forward by an accelerating force in the weightless universe and being stationary in a normal gravitational field, as we sense it on Earth. Einstein concluded that light rays in a gravitational field must experience curvature, and to this day this deflection is a precisely tested prediction of the general relativity.

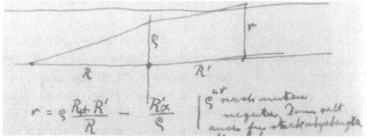

Light deflection sketch from Einstein's notebook (copyright Albert Einstein Archives, Hebrew University of Jerusalem)

Curved light rays reminded Einstein of the principle named after Dutch physicist Huygens, according to which light always seeks the fastest path, not the shortest. This explains, for example, the ability of lenses to deflect and bundle light rays. Thinking about that in 1911, Einstein surmised that in the vicinity of celestial bodies - in gravitational fields, in other words - the speed of light would be slower. This is where the central idea of general relativity is to be found. Clusters of galaxies, for example (which were not discovered until decades later), bundle light in the same way as collecting lenses do. Astronomers today call these gravitational lenses, though without being aware of the direct link to Einstein's idea of the variable speed of light.

THE MYSTERIOUS ROLE OF THE UNIVERSE

Given the subsequent, intricate reformulation of general relativity, it may sound remarkable that its essence can be so easily pictured: under the action of gravity, light moves along a curved path. But it was Einstein's intuitive way of tackling the problem. As happens so often, the first idea was the best.

How the various descriptions fit together will be the subject of the chapters that follow. But the crucial point to be made here is that Einstein's formula of 1911 was a gem he was unable to recognize at the time, even though he had been given a hint of it.

As long ago as 1883, Viennese physicist and philosopher Ernst Mach had published profound thoughts on gravitation that fascinated Einstein throughout his life. Mach had understood that the laws of motion depend on the *relative* movement of bodies only, a fact that Einstein elegantly formulated in 1905 with the theory of special relativity. Furthermore, Mach had also argued that a body's inertial resistance to acceleration, and hence the concept of mass itself, should also depend on the body's motion relative to the rest of the universe. Even remote galaxies would influence the speed at which an apple falls from the tree! This fact would imply that the tininess of the gravitational force has its counterpart in the enormous size of the universe. Einstein had racked his brains about this problem, but he could not find a way of incorporating Mach's idea into his own theory. Unknowingly, in 1911 he had already done so!

Chapter1: The blind spot of physics

The formula for gravitational potential he had developed in his variable speed of light article implied that the gravitational constant itself could be calculated from the mass distribution of the universe – but Einstein didn't notice. The size of the universe couldn't be guessed until the 1930s, after the discoveries of American astronomer Edwin Hubble. At the time there was still discussion of whether there could be other galaxies besides our own Milky Way or not.[i] That means Einstein knew nothing about other galaxies as yet, and accordingly, he had no idea of the true size of the universe.

Nonetheless in 1911, he wrote down a formula that would have excited him, had he been aware of the later data. The mysterious link between gravitational force and the size of the universe, as anticipated by Ernst Mach, would have emerged as the two sides of a mathematical equation. The fact that Hubble's observation came twenty years too late seems to be a tragic accident of the history of science. All of this sounds quite incredible, thus I feel I have to add a series of explanations here.

WHERE WERE ALL THE OTHER PHYSICISTS?

The intriguing connection between the newly discovered size of the universe and the strength of gravity had struck a few researchers in the 1930s - among them Sir Arthur Eddington (who brought Einstein to fame with his legendary solar eclipse expedition) and Paul Dirac, the brilliant co-founder of quantum mechanics and Nobel laureate of 1933.

But neither Eddington nor Dirac saw the link to Einstein's theory of variable speed of light. It took until 1957, two years after Einstein's death, until Robert Dicke, an astrophysicist from Princeton, eventually solved the puzzle. Dicke later became

---

[i] Cf. The famous "Great debate": Curtis-Shapley.

famous for his role in the discovery of the cosmic microwave background, and only bad luck prevented him from winning the Nobel Prize. Dicke, who had studied Ernst Mach, saw the power of Einstein's formula - and improved it in one crucial respect.

For reasons that only Dicke himself could have explained (probably due to an incorrect calculation^2 in his 1957 paper), he abandoned this path. He turned to another theory, which became more prominent, even though it watered down the essential idea. The predictions made by this Brans-Dicke theory failed over the years, which Dicke must have found frustrating. This evidently led him to overlook the value of Einstein's old idea he had rediscovered. Regrettably, Dicke never returned to the key point in his work of 1957. An unfortunate title dealt the final blow, and this pearl succumbed once again to oblivion.

SEDUCED BY GEOMETRY

The question arises as to why Einstein himself didn't take up his stroke of genius after becoming aware of those cosmological observations in the 1930s. However, we may answer that question plausibly, if we take a closer look at the development of general relativity.

Einstein had completed his variable speed of light paper in Prague in June 1911. In the following year, he returned to Zurich, where he met his old college friend Marcel Grossmann, whose father had helped Einstein to get his first job at the patent office in Berne.

Marcel Grossmann was by then Professor of Descriptive Geometry at the ETH Zurich.[i] Curved light rays immediately reminded him of differential geometry, his field of expertise that dealt with curved paths in curved spaces. Einstein was fascinated to learn that his thoughts could be formulated in what for

---

[i] Swiss Federal Institute of Technology.

Chapter1: The blind spot of physics

him at the time was a new math. However, the physics of variable speed of light began to fade. In retrospect, we must probably conclude that the collaboration with Grossmann distracted Einstein from his best idea.

Instead, somehow it became stuck in Einstein's mind that the development of general relativity had been a race of competing versions of formal geometry. In 1913, he and Grossmann published *Outline of a Generalized Theory of Relativity and of a Theory of Gravitation*, which still contained a mathematical inconsistency. They also predicted that the Sun would deflect a light ray by about 0.85 arc-seconds – a figure that later turned out to be wrong.

Einstein had previously obtained the same incorrect value with his variable speed of light theory, because in 1911 he had assumed that the variability originated from a variable lapse of *time* only. He did not realize that *length* scales were also shortened in the gravitational field.

The discrepancy between the wrongly calculated 0.85 arc-seconds and the correct (double) value of 1.7 arc-seconds was always considered by Einstein as *experimentum crucis* deciding between the two *mathematical* versions that he had published in 1913 and 1915. He did not suspect that the correct value came out as well from a variable speed of light assumption, as Robert Dicke showed in 1957. On the contrary: the success of the 1915 formulation was certainly main reason why Einstein himself came to see his brilliant idea of 1911 as a preliminary, misguided attempt.

It is worth emphasizing once again the subtlety here: correcting the prediction from 0.85 to 1.7 arc-seconds was considered by Einstein – and then by the rest of the world – a "breakthrough to the truth"[3], which was justified within the mathematical formulations at hand.

But this has nothing to do with the question of whether general relativity could also (and even better) be formulated with a variable speed of light. Unfortunately its initial version was also incomplete, predicting the lower value of 0.85 arc-seconds. Thus it is easy to confuse it with the incorrect mathematical version of 1913. Virtually all physicists would draw this conclusion at first glance. Hardly anybody is familiar with Robert Dicke's solution to the problem in 1957.

THE BURDEN OF FAME

The formal geometric version of general relativity also became famous by virtue of the dramatic race that Einstein was drawn into with David Hilbert, the eminent mathematician. It was Einstein who first made him aware of the problem in summer 1915. He was (justifiably) afraid that Hilbert's mathematical brilliancy would enable him to snag the discovery. Possibly, Hilbert published the crucial equations earlier, [4] but he was enough of a gentleman to acknowledge Einstein's priority with regard to general relativity as a whole.

Einstein finally achieved worldwide fame in 1919, when Sir Arthur Eddington announced the results of his solar eclipse expeditions in London, where the great and the good of physics had gathered for a meeting of the *Royal Society*. The new theory – incomprehensible to the layman, as it was called – was spectacularly confirmed. Newton was declared dethroned as the scientific sovereign. Who would have thought that Einstein's intuitive idea of 1911 needed just a minor addition[5] to achieve the same thing?

The worldwide recognition that even put Einstein on the title page of the *New York Times* set the geometric formulation in stone for decades, though most people, in a mixture of admiration and incomprehension, gave the details of differential geometry a wide berth. Later, mathematicians rephrased Einstein's approach in a more accessible way, demystifying it to a certain extent. Yet people are still afraid of general relativity because of its mathematical difficulties, though Einstein had already cor-

Chapter1: The blind spot of physics

rectly described the curvature of light rays in 1911 in a very clear manner.

The mathematics Einstein used in 1915 fascinated him in the 1920s as such, and he started working on his "unified field theory", in collaboration with French mathematician Élie Cartan. Around 1930, Einstein was especially enthusiastic about that intriguing attempt to derive the laws of Nature in terms of pure mathematics. Possibly he attached too little importance to cosmology, which was not his principal concern back then.

COSMOLOGY AT THE WRONG TIME

Not even Einstein's trip to Pasadena in 1931, where he met Edwin Hubble (shortly after his sensational findings of the redshifted light), did not change Einstein's state of mind very much. The two researchers' encounter had public appeal, but it just drew Einstein into the debate about whether the universe was static or expanding. Thus Einstein became distracted from the real problems. The questions that Ernst Mach had once brought up faded among the spectacular imagery.

The key factor however was that in the early 1930s, astronomers had only rudimentary estimates of the mass and size of the universe: the data that had provided so much evidence for the variable speed of light theory of 1911. It almost looks as if contact with prestigious science did Einstein the maverick no good at all.

It was not until 1938 that Paul Dirac revealed the mysterious numerical link between the universe and the gravitational constant, which has inspired many physicists ever since. Dirac even discovered an astonishing relationship to the size of elementary particles, which really ought to have excited Einstein. So where was he? And what was he doing?

Einstein's Lost Key

Edwin Hubble and Albert Einstein. In 1931, Einstein visited Hubble at the Mount Wilson telescope in Pasadena. [Emilio Segrè Archives]

It is hard to reconstruct exactly what he was dealing with so long ago. Obviously, he was active in various other theatres of war, such as the unified field theory, the conflict with quantum mechanics ("God does not play dice") - and also, alas, with a very real one: the gloom-and-doom scenario of an atomic bomb developed in Germany, which would have put a disastrous weapon in Hitler's hands. Alarmed by this, Einstein in 1939 signed a letter to US President Roosevelt that helped to start the *Manhattan Project*.

At any rate it is not known - and I have sifted through all the biographies - what Einstein thought about Paul Dirac's observation, or even that he ever heard of it! This highlights the limits of scientific communication at the time – and the looming disaster of World War II cast a long shadow over science, as everything else.

Chapter1: The blind spot of physics

SPLIT WORLDS

Dirac's observation that the size and mass of the elementary particles were linked to those of the universe was as sensational as it was baffling. Obviously it had a great deal to do with Mach's principle, which Einstein admired so much. Even so, it seemed natural to Dirac to assume a constant speed of light in his model - he was completely unaware of Einstein's work published in the *Annalen der Physik* in German in 1911. The reclusive Englishman corresponded little. The two greatest physicists of the 20th Century did not talk to each other about a crucial point of the theory of relativity!

It was probably the missing piece of the jigsaw - the variable speed of light - that led Dirac to make a premature statement about the gravitational constant. He claimed[6] that it must decrease with time, which observations have failed to confirm to this day. Therefore, it is often believed to be ruled out. But even those who know Dirac's work are unaware that there was a direct link to Einstein's theory, as shown - *inter alia* - by Dicke in 1957.

Yet not even Dicke referred explicitly to Einstein's publication, even though the formulae are nearly identical. If Dicke really knew nothing of the 1911 work,[i] this would be remarkable - but in any case further evidence for the idea's huge potential. Dicke was the first to recognize that the theory of variable speed of light describes the solar system as well as the conventional formulation, but it makes new predictions for the universe. The key feature of the idea that goes back to Einstein is that it is equivalent with general relativity for planetary motion, but leads to a dramatically different cosmology.

---

[i] More details follow in Chapter 8.

A DIFFERENT BIG BANG

Dicke gave Einstein's formula a new meaning that one may call revolutionary. While the common interpretation of the Hubble redshift was that galaxies were receding or space was expanding, Dicke drew a simple conclusion from his variable speed of light model. The change in color is merely caused by gravitation that affects the wavelength of the light reaching us from great distances. This interpretation, mentioned almost casually in Dicke's work, would turn our current picture of the universe upside down.

If there were no spatial expansion, there would no longer be a Big Bang in the conventional sense - merely the fact that light propagates. Although redshift as such is undisputed, cosmologists have been struggling for decades to find a correct description of the universe in a much-adjusted model that constantly throws up contradictions. Just imagine: what if the debates about the Big Bang that have continued to this day were based on a single huge misunderstanding, because the expansion of the universe is just an illusion?

This would not only overthrow existing ideas, but would also be a decisive leap in our understanding of the laws of Nature. Modern cosmology is unable to give a *cause* for the expansion of space that has been assumed since Hubble. Dicke, conversely, also provided a *reason* why the universe appears to be *expanding*. Regrettably, Einstein did not witness this interpretation of his formula by Dicke, which would have fitted in so closely with his convictions about physics being simple and comprehensible.

OLD MYSTERIES REAPPEAR

Even if Dicke unfortunately failed to pursue his work of 1957, he still drew physicists' attention to the central problem of cosmology. He was the first to point out the mysterious fact that

Chapter1: The blind spot of physics

universe was what we call "flat". In simple terms, it is surprising that the kinetic energy of all the mass in the universe coincides so precisely with the energy stored in the gravitational field.

But this mystery would be solved by sheer logic if Einstein's 1911 formula were correct, because the two forms of energy would then be equal by definition. Back in the nineteenth Century the great thinker Hermann von Helmholtz pondered the elementary question of why nature had invented two so different forms of energy. Looking at the history of physics, we can see how Einstein's idea of the variable speed of light touches on the most profound, all-but-forgotten problems of the cosmos. Contemporary physics instead 'explains' the flatness of the universe with an exotic theory of 'cosmic inflation', which claims that the universe dramatically ballooned in the first tiny fractions of a second after the Big Bang - postulating fancy concepts such as 'bubbles in a false vacuum', 'branes', 'chaotic inflationary phases' and 'parallel universes'. Einstein adhered to the simplicity of natural laws. Go figure what he would have thought of that.

MODERN MISUNDERSTANDINGS

Modern cosmology has good reason to deal with that idea of Einstein's. The past decades were an era of fantastic observations, but at the same time many of them were poorly understood. This lack of understanding has led to a complicated cosmological model based on a series of *ad hoc* assumptions.

One prominent example is the 'accelerated expansion' of the universe, for which the Nobel Prize was awarded in 2011. The cause of the acceleration is usually attributed to an unknown substance called 'dark energy'. Too bad that the cause of the expansion itself is unknown, but this is generally brushed aside.

'Dark energy' is just a numerical value (around 70 per cent of the mass of the universe) that is fitted to the observations. Einstein was highly suspicious of that sort of thing on principle:[7] "I cannot imagine a rational theory that explicitly contains a number that the Creator could just have chosen differently, had the fancy taken him."

But most of all it would be the content of the data that would astonish him, if he came to know how strongly it supported Dicke's interpretation of his 1911 theory. 'Accelerated' expansion consists primarily of the fact that a slowing down in the expansion, as demanded by conventional theory, has *not* been observed. If Einstein's formula is correct, then the universe is being neither accelerated nor decelerated: it only appears to be expanding. And there would be no need for 'dark energy' of course.

Am I talking too big? It might be objected that alternative scientific theories have to conduct extensive calculations to prove that their results agree with observations. But it is also true that at the root of any scientific revolution, there has always been a succinct concept. Copernicus saw that the planets orbit the Sun. Newton understood that celestial and mundane gravity had the same origin. Thanks to Einstein, we know that while the speed of light does not depend on the observer, it does depend on the position of surrounding masses. This is the central idea, and it has ground-breaking potential.

SCIENTIFIC REVOLUTIONS ARE NOT SCHEDULED

I should admit that the majority of cosmologists would disagree with the arguments put forward here. But equally they should admit that the same majority has not even cast a glance at Einstein's 1911 idea. Until recently the work was hard to access in English,[8] and the scarce citations show that people are mostly unaware of the link to the variable speed of light.

Chapter1: The blind spot of physics

Most physicists have heard *something* about the variable speed of light, of course. But opinion-forming is usually very superficial – partly due to the flood of publications with which science is inundated, forcing the individual to rely on the rapid verdict of authorities.

The erudition of such celebrities is highlighted by one recipient of the Einstein Medal (sic!), who has asserted9 that variable speed of light is senseless and a logical contradiction. Obviously, you can be ignorant about history (he knew nothing of Einstein's article) and win physics prizes. But leaving aside such curiosities, let us take a more general perspective of the problem. Physics has objective foundations, certainly - but convictions in the research communities are merely shaped by sociological processes, which often result in uniform opinions. We might also say: groupthink.

It is the paradigmatic strangeness of Einstein's old idea and its relationship to Mach's principle that renders it the alien thing that is not touched by today's researchers. Too much established wisdom would be under threat, if one would think the unthinkable. Yet such maverick ideas have often turned out to be true in history.

This raises the disconcerting question: who on earth studies articles that are a hundred years old? Pace those involved, research collaborations tend to see the present as all-important. Often there is a lot of money and prestige at stake that makes them churn out the next publication. No prizes are handed out for dwelling on old theories, especially if they threaten to undermine many years' research.

HELP FROM HISTORY

Dealing with past ideas will also look pointless to anyone who considers science as a step-by-step revelation of the truth by means of successive small findings. There is no lack of self-confident physicists who praise the advances in observational cosmology as a precise picture of reality, regarding Einstein - not to mention Mach - as outdated. But what they lack is knowledge of how science works. As philosopher Thomas Kuhn has shown, long periods of 'normal science', characterized by the refinement and growing complexity of the dominant paradigm, usually culminate in a crisis in which the knowledge of generations turns out to be obsolete. It is astronomy that gives us Kuhn's showcase example: the geocentric world view, growing ever more complicated, dominated for no fewer than 1500 years before it was eventually superseded in the Copernican revolution by the simple heliocentric model.

In these cases, existing data has to be reinterpreted under an entirely different paradigm. This does not mean the prediction of some new detail, but starting from scratch, an effort that largely exceeds the capacity of an individual. Thus someone may feel tempted to remark that the present book does not do this. However, I can do little more than remark that current cosmological research is not engaging in any fundamental reflection on Einstein's idea. That will not happen overnight. But cosmology must at least present its results in an impartial and transparent form that does not anticipate the current model.

The unique potential of Einstein's idea lies in calculating the gravitational constant from the data of the universe. All scientific revolutions in physics have found such relations, thus making a constant of nature (or several of them) superfluous: Maxwell's electrodynamics did away with the magnetic field constant, and quantum mechanics got rid of several arbitrary numbers. Einstein's 1911 formula would do the same with the gravitational constant.

This perspective is so fundamental that it must be tested very carefully, even if this or that detail appears to contradict such an interpretation. The connection between the gravitational constant and the universe is probably the sole problem in cosmolo-

Chapter1: The blind spot of physics

gy, but it takes many forms. Physics ought to take the advice of the philosopher Ludwig Wittgenstein:

> *Don't get involved in partial problems, but always take flight to where there is a free view over the whole* single *great problem, even if that view is yet not clear.*

In 1911, Einstein was in the right place. It is now our job to make the view clear.

Einstein's Lost Key

Chapter 2
"I Rely on Intuition"[i]

Why all scientific revolutions began without calculations

> *It is a special blessing to belong among those who can and may devote their best energies to the contemplation and exploration of objective and timeless things – Albert Einstein*

Einstein revolutionized physics. In the first place we have to thank him for discovering the essential laws of nature. But it is also worthwhile to have a look at his working methods that enabled those breakthroughs. This chapter deals with Einstein's way of thinking and arguing, but it is more than a digression. Modern physics operates in such a different style that a present-day student – not to mention an expert – would have trouble recognizing Einstein's intuitive idea of 1911 as fundamental physics. It is, however, the way of reasoning that was practiced back then that led to the revolutionary advances in physics, not the formal calculations that dominate the current paradigm. Although it is an obvious lesson from history, employing simple, clear thoughts takes courage today – because doing so contrasts so overtly with the current fashion. If one fails to understand the intuitive way of thinking practiced by Einstein, it is hard appreciate the significance of his 1911 idea.

---

[i] This phrase was used as the title of a biography (by Abraham Pais), which however barely mentions on the intuitive components of Einstein's insights.

How are laws of nature discovered? As we shall see, there is always a visionary idea at the beginning. What abilities do you need for making such discoveries? This question naturally arises with regard to Einstein's contributions to physics. Few would deny that, in his own way, Einstein was a genius. But if we ask what his genius actually consisted of, opinions will soon differ. Did Einstein simply have more neurons than ordinary mortals? Or were his neurons better connected, thanks to early conditioning or talent? What exactly was it that Einstein's brain could do better than the brains of other people?

Sometimes the impression is created that it was Einstein's ability to carry out intricate calculations that marked him out from the rest. Viewed in perspective, his skills were certainly at a high professional level in comparison with those of other researchers. But they were not outstanding, as Einstein himself frequently admitted. His correspondence with the French mathematician Élie Cartan, for example, is full of apologies for his mathematical clumsiness.

David Hilbert was much more accomplished in calculation, and so were plenty of other physicists. It would be wrong, however, to assume that Einstein felt inferior to them. His modest formulations sometimes hide the fact that he considered his ability to detect the laws of nature to be as unique as it indeed was. Otherwise he would not have stubbornly pursued his ideas for decades, in the teeth of almost all other physicists. He trusted himself and, above all, his intuition.

IN THE BEGINNING WAS THE WORD

But was he right about that? Do we place him on too high a pedestal? I do not think so. Let's take a look at the history of science, which shows us how the crucial breakthroughs for our civilization came about.

Although a law of nature in its final form needs to be expressed in terms of mathematics, the real discoveries originate

Chapter 2: „I Rely on Intuition"

from intuitive thoughts, the seeds of the scientific revolutions that we have seen throughout history. In mathematical terms, Newton's law of gravitation states that when the distance from an attracting mass is doubled, the force decreases to one fourth. But Newton's true flash of genius was in realizing that celestial and earthly gravitation had the same origin. One might object that many people with less mathematical talent might have had this idea, and that Newton only formulated it correctly. But this underestimates the mental toughness with which Newton pursued his idea, as well as the accompanying conceptual achievements that came from intuitive thinking. First Newton had to devise notions such as force, speed and acceleration. Only after this spadework did the inverse-square law fall into his lap like a ripe apple (sic!).[i]

As with celestial and earthly gravity, the creative transfer of a phenomenon to another context is the mental achievement that leads to a great discovery. Take, for example, the revolution in atomic physics. The spark for this was Niels' Bohr's application of a property of light to matter – a leap of imagination back then. Bohr realized that the constant h, discovered by Max Planck,[ii] had the unit of angular momentum ($kg\ m^2/s$), and he related it to a possible orbital motion of electrons in atoms. His reasoning was devoid of mathematics, an intuitive association, like making up a movie in one's brain. Bohr was even less well-versed in calculation than Einstein, which Pauli and Heisenberg often used to tease him about. Even so, Bohr discovered more.

Another result of Bohr's intuition is even less well known. It was Bohr who realized that chemical properties depend on the number of electric charges in the atomic nucleus. Bohr thus

---

[i] Copernicus, whose contribution may have been the most important, formulated his theory without any mathematics.
[ii] The famous formula for the energy of light quanta $E=hf$ had been found by Einstein, in opposition to Planck's convictions.

completed the work of Mendeleev, and in one respect even the picture of atoms expressed by Democritus. Today these seem almost trivial, but it is typical that in retrospect we barely appreciate such ingenious associations. They are so obviously correct that they do not need a mathematical justification: they are facts found by pure intuition. Bohr's discoveries also demonstrate that there is no need to be ashamed of an idea just because it looks simple.

TRUTH AT FIRST SIGHT

There are lots more examples, such as the origin of thermodynamics, an essential field of physics. The German physicist Robert Mayer (1814-1878) discovered that temperature is nothing other than the kinetic energy of particles – that was the essential, intuitive idea. Mayer was so bad at arithmetic (I take this example on purpose) that he derived the formula for kinetic energy, $\frac{1}{2} mv^2$, with an incorrect numerical factor. Nevertheless, his achievement remains unique.

Science has made big leaps as a consequence of such qualitative thinking: thinking about constants of nature and their units, through visual association or creative transfer. Usually these groundbreaking insights are given far too little respect because in retrospect they appear so simple.

Ernst Mach's intuitive conjecture that the weakness of gravity is related to the size of the universe also falls into this category, as does Paul Dirac's hypothesis that the size and mass of elementary particles determine the strength of their interaction. We shall have more to say about both Mach and Dirac, because their visions of unification have not yet been realized. And of course, Einstein's idea of the variable speed of light is one of these brilliant intuitions.

> *The supreme task of the physicist is to arrive at those universal elementary laws from which the cosmos can be built up by pure deduction. There*

Chapter 2: „I Rely on Intuition"

> *is no logical path to these laws; only intuition, resting on sympathetic understanding of experience, can reach them* – Albert Einstein

Einstein saw a link between gravity and the properties of light, and suggested that variable speed of light was responsible for the curvature of light rays. The idea can be grasped visually, with no mathematics, but it implies nothing less than the unification of gravitation and optics.

Einstein's previous reflections that led him to the assumption of curved light were also purely intuitive. He sensed – there's no better way of putting it – that inertial mass, which resists acceleration, and gravitational mass, which creates attraction, had to be identical. We shall look more closely at this so-called principle of equivalence in the fourth chapter.

To summarize, the history of science provides plentiful evidence that crucial findings, from Einstein to Bohr and back to Newton and Copernicus, sprang from intuition. But the question remains: what does intuitive thinking actually mean?

AN UNDERRATED FORM OF THINKING

> *I have the result, I just don't know yet how I'm going to get there!*[10]

Einstein could not deal scientifically with brain research in his own time. Psychology as such was founded by his contemporary Sigmund Freud, and half a century was to pass before cognitive psychology became an empirical science. Today we know a little more about the subject, and its results also justify Einstein's approach to dealing with the laws of nature.

The complexity of the human brain hosts very different working modes, each of them responsible for amazing capacities. The theory of the "left" and "right" parts of the brain, which have astonishing functional differences, has become relatively well known. According to these findings, the left half of the brain acts more algorithmically and logically, while the right half prefers a comprehensive view and works in a concrete and intuitive manner. The topic has recently been discussed in Daniel Kahnemann's bestseller *Thinking Fast and Slow*. Kahnemann has shown that the "first", intuitive, system often works several orders of magnitude faster, more safely and more efficiently than the abstract, slow system that guides mathematical processes.

Einstein's summer house in Caputh near Berlin, a retreat for sailing and thinking.[Stephan M. Höhne, GNU license for free documentation]

If we consider the evolution of Homo sapiens, it is quite clear why some mindsets suit us more and others less. The development of the brain that ultimately distinguishes us from our animal relatives took place in an environment in which quick, heuristics-based decisions were a matter of life and death. Three-dimensional imagination and the ability to preprocess our movements in spatiotemporal simulations are the outstanding characteristics of our brain. Einstein made particular use of them in his thought experiments.

In contrast, little evolutionary advantage has accrued from the ability to estimate the logical consistency of lengthy formulae,

Chapter 2: „I Rely on Intuition"

or from performing abstract calculations at high speed. In ancient times the saber-toothed tiger would have had us for dinner, and today the computer shows us how embarrassingly slow the human brain is at "dull" arithmetic.

DRIFTING OFF INTO THE UNIVERSE OF HOMO MATHEMATICUS

Could it be that the greatest discoveries that humanity has elicited from nature were just the fruits of the most sophisticated mathematical tools? I doubt it. The physical laws found in modern times require calculations that go well beyond everyday math, but certainly an unlimited number of researchers can acquire these skills. Differential equations and wave functions in quantum mechanics are not really difficult (though the underlying concepts are), and even the general relativity – which is often depicted as impenetrable – can, if properly taught, be understood by any student. We shall however come back to the question of whether the usual representation is really the simplest one.

But why are the methods of contemporary physics so different? For anyone who opens a copy of the *Physical Review* it becomes evident that current theoretical physics primarily deals with massaging mathematical formalisms. It is, however, also evident how drastically this style differs from what could be found in physics journals fifty or a hundred years ago. And one thing is even more obvious: how few fundamental discoveries physics has made since then.

It is often argued that the easy laws of nature are already discovered, and that everything else can only be accessed by arcane mathematics. But that's not why present-day physics is in crisis. The crisis has arisen because the culture of conceptual reflections – as fostered by visionaries like Mach, Einstein, Schrödinger, Bohr, and Dirac – has gone missing.

Einstein's Lost Key

The good old days are – sadly, sadly – over, today's theorists claim, bemoaning the fact that the era in which you could discover something by simple calculations has ended. In particular, string theorists gloss over the obvious lack of success of their excessive calculations by resorting to these arguments. But there is no justification for this distortion of history apart from their own inability to think clearly and intuitively.

PURE CALCULATIONS LEAD NOWHERE

> *Although I am a typical loner in my daily life, my awareness of belonging to the invisible community of those who strive for truth, beauty, and justice has prevented me from feelings of isolation –*
> *Albert Einstein*

The view that fundamental discoveries need new mathematical worlds is widespread among contemporary physicists, and unfortunately it is widely parroted. While unsupported by facts, the origins of this fashion can be traced back almost a hundred years, to when Felix Klein, David Hilbert and Hermann Minkowski in Göttingen mused over whether new physical discoveries were perhaps only to be found in mathematics.

Minkowski, in 1908, had introduced the notion that three spatial dimensions plus time would form four dimensions, thus giving an excessively formal interpretation to Einstein's special theory of relativity. Minkowski indulged in euphoric statements about the "real" four-dimensionality of the world, failing to see that space and time are in fact quite different things. "Since the mathematicians have invaded the theory of relativity", Einstein remarked mockingly, "I do not understand it myself anymore." Minkowski's ambitious interpretation turned out to be disastrous in the ensuing period, because it hindered a deeper reflection on the concepts of space and time.

Chapter 2: „I Rely on Intuition"

Lumping together such different phenomena as space and time is a conceptual absurdity that makes a mockery of genuine understanding. But precisely this fashion of formal reasoning was established back then, and led physicists to misinterpret the speed of light c as a mere "conversion factor" for turning space into time. It is clear that excising the meaning of a constant of nature must have nonsensical consequences. But since then the formal mathematical approach has triumphed over the intuitive.

The paradigm embarked on at that time continued in post-war physics: basic assumptions about nature became more and more bizarre, and it was increasingly obvious that the theories had lost touch with reality. In my books *Bankrupting Physics* and *The Higgs Fake* I have described in detail how the method of finding laws of nature through excessive calculations has taken on a life of its own. Ultimately this is an issue that sociology will have to address in due course. But any reader who has become familiar with Einstein's 1911 idea will feel that modern fancies such as "string theory", "cosmic inflation" and "multiverses" have nothing to do with reality.

> *I have no special talents. I am only passionately curious – Albert Einstein*

EINSTEIN WAS A LITTLE GUILTY, TOO

I shall not attempt to deny it: my argument that Einstein's approach to physics was intuitive puts me in something of a quandary. In the years after 1919, Einstein himself utilized purely mathematical structures in his search for a "unified field theory". Many people criticized him for that, too.

Three things ought to be said here. First, Einstein always struggled to bring his calculations back to phenomenology. Consider, for example, the last section of his 1930 paper,[11] where he emphasized that calculating the masses of the electron and the proton ought to be the next step – a puzzle that remains unsolved to this day.

Secondly, when dealing with differential geometry, Einstein let himself be guided by his intuitive mindset, and he uncovered amazing insights. Nothing testifies better to this than the correspondence[12] between Einstein and the French mathematician Élie Cartan, in which Einstein's ideas always played the creative part.

Thirdly, Einstein – and probably many of his followers – was seduced by the elegance that physical laws necessarily contain at this level of abstraction. He wrestled with the geometric difficulties, and without the help of Marcel Grossmann he could hardly have made so much progress. However, the unified field theory was not the right way. This is not easy for me to admit, because the Einstein-Cartan theory fascinated me for a long time. Nevertheless, this search into formal constructions not only led to very little progress, as most people would agree, but also prevented Einstein's more direct, physical ideas from developing.

"THE LORD CREATED THE DONKEY AND GAVE HIM A THICK SKIN."

Unthinking respect for authority is truth's greatest enemy. – Albert Einstein

Maybe it was the contact with colleagues with great mathematical skill that diverted Einstein from the path of intuitive physical discoveries. Intuition is a very individual experience, often difficult to put into words, and almost impossible to translate into a formal line of reasoning. It is certainly not a coincidence that revolutionary ideas in physics always originated in flashes of inspiration.

Intuition is strongly related to critical thinking. Einstein's intuitive insights were combined with mental strength to hold on to his ideas. Even when he was surrounded by opposing convictions, as in the case of his light-quantum hypothesis, he fearless-

Chapter 2: „I Rely on Intuition"

ly challenged the authority of Max Planck. Einstein's obstinacy was certainly a decisive feature of his success. In the last three decades of his life, he also fought a lone battle against the prevailing interpretation of quantum mechanics. His argument "God does not play dice" has been criticized, but in a general sense he was right to demand explanations rather than a purely descriptive theory.

Einstein's healthy self-assurance, albeit cloaked in polite words, was evident from the time he was a graduate student,[i] if not earlier. A professor once advised him against the study of physics, since he feared it would be too demanding for Einstein's mind. There were other faculties, after all, he said. "Professor", Einstein responded, "as regards the other faculties, I have even less talent." Einstein had learned early not to bow to authority. His seemingly self-critical, but highly ironic, description of a "good student" is also very illuminating:

> "In order to be a good student, you need easy comprehension; willingness to focus all your attention to what is presented; orderliness, for recording what is told in the lectures and elaborating it conscientiously. To my regret, I realized that all these characteristics were totally lacking in me. Little by little, I learned to live in peace with my somewhat guilty feelings and organized my studies in accordance with my intellectual stomach and my interests. There were a few lectures I followed with great interest but I also skipped many. At home however, I studied the masters of theoretical physics with zealous enthusiasm. This was good in itself, and it served to mitigate the bad conscience so effectively that my

---

[i] His professors thought him too independent and unconventional to be considered for an assistant's position that he was interested in at the ETH in Zurich.

> *psychological balance was not disturbed in any way."*

Here, too, Einstein had managed to escape from sociological pressure – he cultivated his independent, intuitive thinking from an early age. At the age of only 17, he had imagined riding on a light wave, thus preparing the world of ideas that later led him to the special theory of relativity.

This chapter has dealt more with Einstein's working methods in general than with his specific idea of variable speed of light. This was not primarily the result of biographical interest. Great achievements are inseparable from the way in which they come about, and contemplating Einstein's personality as a whole therefore helps us to judge the value of his discoveries in the context of modern science, which is so different from that of earlier times. Einstein's distinct individualism – which could stretch to stubbornness – was part of a working method guided by intuition. But it was not only the revolutions in physics triggered by Einstein that were based on intuitive ideas: so was practically every breakthrough in the history of science.

> *Few of us are capable of calmly expressing opinions that dissent from the prejudices of their environment; most people are actually incapable of ever reaching such opinions – Albert Einstein*

Chapter 3
"Arbitrary Numbers ... Ought Not to Exist."

Einstein on simplicity in nature

The primary objective of this book is to introduce you to Einstein's idea of variable speed of light. His ingenious approach touches on problems of physics that are even more fundamental than the theory of relativity, but in order to grasp their importance, we must take another deviation on method and consider the role of the constants of nature in physics. Nature reveals these messages when we are measuring, and throughout the history of physics they have always been the royal road to profound understanding.

At the genesis of a scientific revolution, there is always a conceptual idea. This is usually accompanied by the discovery or explanation of one of those mysterious numbers that seem to determine the laws of nature everywhere in the universe. The question as to the origin of these constants of nature extends into the realm of natural philosophy. Only by pondering it can we properly value Einstein's idea of variable speed of light. Einstein did reflect on such general questions about the constants of nature, but, surprisingly, the thoughts on this subject of the greatest scientist of the modern age have remained largely unknown.

Certainly Einstein failed to make "official" statements about these matters in his writing. We are therefore lucky that, around 1920, Einstein came across the physics student Ilse Rosenthal-Schneider. Thanks to her correspondence with Einstein, we now have access to his deep convictions about nature, which otherwise would have been lost forever.

THE PHILOSOPHER

In 1920, Ilse Rosenthal-Schneider was one of the first women to be awarded a PhD in philosophy from the University of Berlin. She probably first met Einstein in 1919, and she was the interviewer who asked him what would happen if observations were to contradict the general theory of relativity, teasing out Einstein's famous answer: "Then I'd feel sorry for the good Lord. The theory is correct." Maybe Einstein's words, taken so seriously, were also a form of flirtation with a woman who, at 28, was twelve years younger than him. They often held long conversations on the tram after Einstein's lectures, but it is also clear that Rosenthal-Schneider was not one of Einstein's lovers during the Berlin period; these have been thoroughly researched by biographers.

What we are dealing with here, instead, is a correspondence that Einstein conducted with her, with obvious appreciation for her questions. Rosenthal-Schneider had fled Nazi Germany in 1938, and after World War II she settled in Australia. The Nobel Laureates Max von Laue and Max Planck also addressed her questions about the fundamentals of physics in detail. It almost seems that this modest woman disguised her highly significant thoughts as questions to these celebrities, keeping her part of the discussion out of the limelight. Incidentally, she summarized her own letters only in indirect speech, and several intelligent ideas can only be identified from the responses. We shall come back to this later.[13]

In order to appreciate Einstein's 1911 idea, we must also reflect on the very existence of the fundamental constants of nature. Why are they needed at all? Can we imagine physics without constants of nature, and if not, why not? Unfortunately, back then in 1911, Einstein had not yet said anything about

Chapter 3: „Arbitrary numbers ... ought not to exist"

constants of nature in general, and this was one reason why he was unable to recognize the full potential of his idea.

Even so, the importance of the variable speed of light that I intend to demonstrate fits in exactly with Einstein's beliefs, even if we came to know these beliefs much later, indirectly and only by chance. This is Rosenthal-Schneider's merit; in an exchange of letters in 1945, she made Einstein comment on the issue of constants of nature.[i] The discussion started with dimensionless physical constants, that is, pure numbers that are derived from combinations of measuring values. After initial misunderstandings, Einstein wrote to her as follows on 13 October:

> "I cannot imagine a unified and reasonable theory which explicitly contains a number that the whim of the Creator might just as well have chosen differently." [ii]

Later, on 24 March 1950 he reiterated this in a slightly different context:

> "If dimensionless constants in the laws of nature could from a rational point of view have other values as well, they shouldn't exist. My 'faith in God' leads me to see this as obvious, though there may not be many who share this opinion."

Einstein later described this as "not categorical assertions, merely suppositions based on intuition". This does not, of course, mean that he was not entirely convinced.

One may also formulate Einstein's statement like this: 'Why should nature assign a special significance to an arbitrary, incal-

---

[i] Rosenthal-Schneider remarks about this: "In contrast to his ... nonreaction when asked about his health or other matters, Einstein replied immediately. He only did that with regard to scientific problems."
[ii] Although Einstein was an atheist, he often used the idea of God in his arguments when he was intuitively convinced of something.

culable number?' Evidently it would be irrational thinking, a thing he disdained. Indeed, no true scientist can be satisfied with being 'served up' unsubstantiated numbers by nature. He wants to understand and calculate them. Accepting unexplained numbers as 'constants of nature' would be a regression to pre-scientific thought, similar to primitive civilizations who put unexplained phenomena down to the whim of the gods.

PHYSICS MEANS THINKING ABOUT CONSTANTS OF NATURE

No scientist should therefore feel the slightest unease in identifying with Einstein's statement, but it is remarkable that he formulated it so early. In 1950, only a few of these numbers were known. Presumably, Einstein had in mind the so-called fine structure constant, the reciprocal of which had been determined at the time as approximately 137. Since then, the measurements have been improved to $1/\alpha = 137.03599976$, yet the origin of this number remains unknown to this day.

The fine structure constant is just one of several examples, but the one for which all great physicists have stepped in to question this kind of numerical value. Paul Dirac, whose ideas will be a major theme of this book, scorned all attempts to sweep this puzzle under the carpet. Whenever young theoreticians came up to him with new ideas, he used to ask about the origin of this number – and brushed them off if they had not thought about it.

Richard Feynman, the Nobel Laureate of 1965 and an icon of post-war physics, wrote:[14] "All good theoretical physicists put this number up on their wall and worry about it." Einstein was therefore by no means the only physicist to take such a view: he was just the first to express it in general terms.

We might formulate this 'working philosophy' as follows: it is the physicist's job to explain, not merely to describe. No critical mind will ever be satisfied with the idea that nature blessed

Chapter 3: „Arbitrary numbers ... ought not to exist"

us with numbers like *137.035999...*, numbers that cannot be calculated as a matter of principle. Elementary logic makes you suspect that there is something that we haven't yet grasped, and it would surely be hubristic to exclude that possibility.

The reason why I am presenting these arguments in abundance is that present-day physics, alas, has more or less bidden farewell to the principles of economy of thought. Cosmology and particle physics now accept dozens (!) of unexplained numbers in their so-called "standard models". Since I have already described these developments in detail,[i] we now turn again to the rational physics of Einstein.

> *It is at the heart of the quest for knowledge to seek simplicity and parsimony of the basic hypotheses – Albert Einstein*

EVEN MORE NUMBERS, EVEN MORE PUZZLES

$\alpha = 1/137.035999 = \dfrac{e^2}{2hc\varepsilon_0}$ is a constant that is derived from other constants of nature: the speed of light $c = 299{,}792{,}458$ m/s, Planck's quantum of action $h = 6.626 \cdot 10^{-34}$ VAs², the elementary charge $e = 1.602 \cdot 10^{-19}$ As, and the dielectric constant $\varepsilon_0 = 8.85 \cdot 10^{-12}$ As/Vm. The above combination is unique because all units such as meters, seconds, amperes and volts cancel out,[ii] so all that remains is a pure number, called 'dimensionless'. Obviously *1/137* does not depend on our definitions of meters, amps etc., which are, after all, arbitrary. Other intelligent civilizations, if they exist in the universe, presumably also measure this number (albeit not with the decimal system). Einstein refers to such dimensionless numbers in his argument, but

---

[i] See my books *Bankrupting Physics* and *The Higgs Fake*.
[ii] With the correlation VAs= Nm and N=kg m/s².

his underlying conviction was not to accept such unexplained, seemingly arbitrary, dictates of nature.

The situation is not really different if we contemplate the 'dimensionful' constants of nature such as c, h, G or the elementary charge e. Alien civilizations would probably not use the same definitions for the elementary charge, but they would surely also stumble across the fact that charge only occurs in small packets. The same applies to the constant h.[i] Here, too, they would have discovered that light only emits energy in quanta of the amount $E=h\cdot f$. Why, oh why? And why is $h=6.626 \cdot 10^{-34}$ so small? Einstein worried about these questions.

One fundamental constant that will play a central part in our deliberations is the gravitational constant: $G=6.673\cdot 10^{-11}\,\frac{m^3}{s^2\,kg}$. It reflects a specific characteristic of nature – the power of gravitation. G was first determined in 1798, almost one hundred years after its importance had been revealed by Newton.

PHYSICAL UNITS – A KEY TO UNDERSTANDING

One may therefore categorize 'dimensionless' constants of nature (numbers such as 137...) and 'dimensionful' constants of nature such as c, h and G – but this does not mean that the latter do not have fundamental significance as well. The "dimensions" of meters and seconds are needed simply because the speed of light – like any other speed – is expressed in the physical unit meters per second. Obviously, the origins of the definitions of meters and seconds go back to the properties of our planet, and

---

[i] It is irrelevant here that the speed of light c is now defined as 299,792,458 m/s. Of course, this definition does not contradict Einstein's idea of a variable speed of light, because if meters and seconds change simultaneously, the ratio between them stays the same.

Chapter 3: „Arbitrary numbers ... ought not to exist"

thus other civilizations in the universe would use a different set of numbers. Anyway, the existence of a constant of nature as such does mean something. The very fact that a limiting speed (applying to all masses, according to Einstein's theory of relativity) exists in the universe remains a remarkable phenomenon.

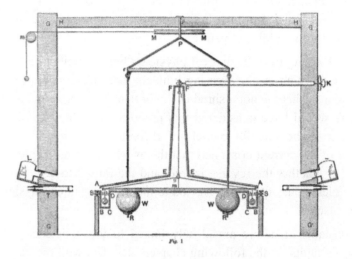

Henry Cavendish's torsion balance (after an idea by Michell) was used for the first measurement of G.

Why does modern physics hardly discuss these questions at all? What does it really mean that we are faced by "constants of nature" – numbers and physical units that come to us without any further explanation? Is nature a dictator, whose quirks we are supposed to accept? Is it sacrilege to ask where the constants of nature come from? Are we to be blamed for doing so?

I don't think so. History suggests that we have simply not yet understood the origin of many constants of nature. There is no serious reflection on these elementary problems among today's researchers. And the more time passes, the bigger the collaborations grow, the more money is pumped into the current paradigm – and the less likely it becomes that such reflection will take place.

BLOODLESS PHYSICS AND MODERN DISEASES

Modern physics at a theoretical level literally disregards the problem. Undergraduate students are already told to do calculations without units, by setting constants such as the speed of light c, Planck's h and the dielectric constant ε_0 to 1. It may help algebraic number fiddling, but it does certainly not invoke sensitivity for the riddles of nature.

Even worse, most theoretical physicists believe units to be irrelevant, which is like wearing blinders that hide fundamental questions. There is not a shred of doubt that Einstein, Dirac and others would have considered this preposterous. One can only wonder if the fanciful models of theoretical physics are the cause of the current crisis and that this is why units are neglected – or whether the reverse is true, and the failure to deal with such elementary questions has led to the mess in which physics finds itself.

Constants of nature are closely linked to the 'existence' of physical units. In the following chapters that deal with the basic properties of space, time and matter, we shall have a closer look to the units of meter, second and kilogram. They are directly related to the constants of nature h, c and G.

As Max Planck first noted, the meter is derived from $\sqrt{\frac{Gh}{c^3}}$, the second from $\sqrt{\frac{Gh}{c^5}}$ and the kilogram from $\sqrt{\frac{hc}{G}}$, albeit with irritatingly small values such as 10^{-35} m. These tiny values have no particular significance and result merely from a lack of understanding. This will become clear in Chapter 9, when we consider the coincidences found by Paul Dirac. However, it is important that the three concepts of length, time and mass are connected with three constants of nature. While being ignorant of the work of Dirac, many fields of modern theoretical physics

Chapter 3: „Arbitrary numbers ... ought not to exist"

(such as string theory and cosmic inflation) try to impute a meaning to the so-called Planck scale. Since values such as 10^{-35} m or 10^{-43} s are far beyond experimental feasibility, they are particularly suited to be a playground for theoretical fantasies. If Einstein's 1911 idea is correct, then it would become obvious that the Planck scale is a misunderstanding had misled generations of theorists.

On the other hand, the fact that a meter could always be expressed as a multiple of the "Planck length" is often misleadingly used as an argument that physical units are irrelevant. This is nonsense. Dimensionful constants such as the speed of light c, Planck's quantum h and the gravitational constant G are simply not "meaningless". I have to reiterate this because theoretical physics is thoroughly infested with this ideology. If we were to concede that the advocates of this view are correct, then not only would Einstein's attempts that are being considered here be pointless, but our entire civilization would look quite different.

Electromagnetic waves – the basis of wireless communication – might well not have been discovered by Heinrich Hertz in 1888 if Wilhelm Weber[i] had not suspected the spectacular connection $1/c^2 = \varepsilon_0 \mu_0$ that links the speed of light with the electromagnetic constants.[ii] Weber noticed that the unit of speed (meters per second) was contained in the product $\varepsilon_0 \mu_0$ and could be measured in the laboratory. A Wilhelm Weber from an alien civilization would certainly have used different symbols, but if we ever communicate with such a civilization, then there will have been someone there who has pondered on constants of nature and their physical units. Physicists who seriously claim

---

[i] A German physicist and close friend of Carl Friedrich Gauss. The unit of magnetic flux (Vs) is named after Weber.
[ii] Historically the notions were different. We use the more didactic approach.

that dimensionful constants are meaningless are delivering their own verdict.

> *The prevalence of fools is insuperable and assured forever. The terror of their tyranny is ameliorated however by a lack of consequence* – Albert Einstein

ESSENTIAL MESSAGES OF NATURE

The example of $1/c^2 = \varepsilon_0 \mu_0$ was chosen on purpose because, in a similar manner, a spectacular connection between constants of nature could relate to the gravitational constant G. Einstein would probably have discovered it in 1911, had he been aware of the true size of the universe (just as Wilhelm Weber was aware of the order of magnitude of the constants involved). G would thus have become superfluous!

This also leads us to a key point of Einstein's philosophy that is outlined in his correspondence with Ilse Rosenthal-Schneider. Laws of nature have to be simple in order to be credible. Simplification in turn means reducing the number of constants of nature that we do not understand.

If one was able to *calculate* the fine-structure constant or work out another arithmetical relationship between the constants of nature, physics would in each case become a great deal simpler. This is where the discovery lies: the number of arbitrary parameters is reduced and our world view becomes simpler. The fewer unexplained numbers nature tells us, the more we have understood.

> *A theoretical construct has little prospect of being true if it is not logically very simple*[15] – *Albert Einstein*

Chapter 3: „Arbitrary numbers ... ought not to exist"

IMPORTANT PHYSICAL CONSTANTS

Name		Value	Unit	Obsolet due to
Speed of light	c	299792458	m/s	
Quantum of action	h	$6,626 \cdot 10^{-34}$	kg m² /s	
Newton's constant	G	$6,673 \cdot 10^{-11}$	m³/s² kg	Epicycles
Elementary charge	e	$1,602 \cdot 10^{-19}$	As	$\alpha = \dfrac{e^2}{2hc\varepsilon_0}$
Dielectricity	ε_0	$8,8542 \cdot 10^{-12}$	As/Vm	$1/c^2 = \varepsilon_0 \mu_0$
Permeability vacuum	μ_0	$4\pi \cdot 10^{-7}$	Vs/Am	Def. Strom
Fine-structure c.	A	$1/137{,}03599$	–	
Boltzmann's c.	K	$1{,}38 \cdot 10^{-23}$	J/K	$kT = \tfrac{1}{2} mv^2$
Rydberg constant	R	$1{,}097 \cdot 10^7$	1/m	$R = \dfrac{m_e e^4}{8\varepsilon_0^2 h^3 c}$

Overview of constants of nature and their respective relations (incomplete). A constant of nature becomes superfluous whenever it can be expressed by others. This has often had revolutionary consequences.

EVIDENCE FROM HISTORY

Weber's visionary discovery expressed in the formula $1/c^2 = \varepsilon_0 \mu_0$ is just a very obvious example: once the equation was established, three constants of nature were turned into two – this was the revolutionary progress. Postulating simplicity, as Einstein did, is thus not an ideology. History, which in this case is an observational science as well, provides evidence that simple theories are correct. Weber and Maxwell surmised that there was a formula, but the big revelation was that light turned out to be an electromagnetic wave. Their visionary ideas led to the unification of optics and electromagnetism.

The potential of Einstein's idea is on the same level: a formula connecting the gravitational constant with the speed of light would unify optics and gravity.

Einstein's Lost Key

If one follows this idea consistently, no unexplained constants at all should survive – and indeed, it was Einstein who suggested this final consequence in the aforementioned letters. Einstein's working philosophy based on simplicity (if we can call it this) should have an obvious appeal for every thinking researcher unspoiled by contemporary "physics". It is close to a logical necessity for anyone who has engaged in reflections about nature that go beyond formulae and data. Real breakthroughs, as history has shown many times, always go hand in hand with simplification.

In the previous chapter it became clear that most scientific revolutions were triggered by an intuitive idea dealing with some kind of unification. If such a surprising connection exists, the second step usually consisted of expressing the idea in a formula that involves a fundamental constant of nature. Either an existing constant was explained, or a new constant was discovered.

Robert Mayer's discovery of the connection between particle motion and temperature led to the unification of mechanics and thermodynamics, which was later completed by James Prescott Joule. The conceptual idea is then shaped by a formula that links kinetic energy to temperature T: $½ mv^2 = kT$. At the same time, this formula represents the 'discovery' of the natural constant k (the Stefan-Boltzmann constant – it makes no difference that its numerical value, $1.38 \cdot 10^{-23}$ J/K, was not determined until later). The quantitative connection, which so impressively confirms Mayer's initial idea, is embedded in k.

Is k an important constant of nature or a trivial definition? Physicists often engage in a heated debate about this, which indicates that they do not have a clear idea about epistemology. Because it depends! Once a law of nature is absolutely clear and beyond doubt, we do not perceive it as such any longer. We *could* argue from a present-day perspective that k is not a "real" natural constant but just a definition of temperature, a unit that

Chapter 3: „Arbitrary numbers ... ought not to exist"

has also been demoted to secondary importance. It has become just a name for the average kinetic energy of a particle. But precisely this simplification or trivialization of the laws of nature was the great achievement of Mayer and Joule.

> *Any idiot can make things bigger, more complex. It takes an added touch of genius – and a lot of courage – to move in the opposite direction – Albert Einstein*

REVOLUTIONS STEP BY STEP

The same pattern appears in the case of Newton's law of gravity. Today the Earth's gravity, $g=9.81$ m/s^2, seems one of our planet's trivial characteristics. But determining this acceleration was a masterpiece that required a Galileo, who obtained the value after a series of measurements and intelligent abstractions.[i]

Newton's conceptual unification of earthly and celestial gravity found its mathematical expression in the formula $g = \frac{GM}{R_e^2}$. This combines Galileo's g with Newton's gravitational constant G. Once the Earth's mass M and its radius R_e are inserted, g becomes obsolete. Therein lies the revolutionary simplification.[ii]

Einstein was dealing with just such a revolutionary simplification in 1911. It is expressed in a simple formula that we shall consider in Chapter 7. Its fundamental importance lies in the fact that the gravitational constant G became superfluous – or "trivial", if you like. Einstein had gone beyond Newton's discoveries.

---

[i] Geophysicists honor Galileo with the unit of acceleration: 1 Gal = 0.01 m/s^2.
[ii] Historically, it would be more correct to say there is a relation to the product GM (before the Cavendish experiment).

Einstein's Lost Key

IMPORTANT UNIFICATIONS IN PHYSICS

Unification		Year	Formula	obsolete
Mechanics and Thermodynamics	Mayer, Joule	1841	½ $mv^2 = kT$	k
Electricity and Magnetism	Maxwell	1865	Maxwell's equations	μ_0
Electromagnetism and Optics	Hertz, Weber	1888	$1/c^2 = \varepsilon_0\mu_0$	ε_0
Dynamics and Gravitation	Newton	1687	$g = \frac{GM}{r^2}$	g
Mechanics and Optics	Einstein, Dicke	1911 1957	$G = c^2/\sum_i \frac{m_i}{r_i}$	G

Overview of revolutionary unifications in physics. These regularly led to one constant of nature becoming superfluous.

What Einstein did in 1911 would thus have squared entirely with the conviction that he formulated much later: the laws of nature have to be simple. Simplicity can only mean getting along with the smallest possible number of arbitrary constants of nature. Newton's gravitational constant G, although it brought about a dramatic simplification itself, nevertheless remained such an arbitrary number. To make it redundant would have meant the unification of optics and gravitation. In 1911 Einstein was on the verge of that revolution. The concept is analogous to Newton's idea of linking the gravity of the Earth with that of the solar system: likewise, the gravity of the solar system can be linked to that of a larger structure, the whole universe. It was the Viennese philosopher Ernst Mach who made the crucial contribution to this fantastic idea, and we shall have a closer look at it now.

Chapter 4

A Precursor of Deep Thoughts

Relativity, Mach and the secret of the universe

Now let us consider the basic idea that led Einstein to the general theory of relativity. It grew out of an intuitive reflection that he first formulated[16] in 1907. Before the relation to Ernst Mach's ideas is pointed out, we must briefly look back at Einstein's special theory of relativity. Today every physicist is so familiar with it that the huge conceptual achievement that lies behind a few simple formulae is almost hidden. Newton's laws of motion had been deemed correct for over two hundred years, and the finiteness of the speed of light, discovered by astronomer Ole Rømer in 1676[i], was not considered a problem for Newton's theory.

Yet the speed of light was not just an isolated curiosity of nature: it had a profound significance for material bodies. Nothing can move faster than the speed of light, which became clear thanks to Einstein's theory. Galileo[17] had understood that the laws of nature do not depend on the motion of an observer, and Einstein applied this insight to the speed of light, c. He realized that c did not depend on the observer: the beam coming out of the headlight of a moving train has always the same speed, irrespective of whether it is observed from the platform or from the

---

[i] From the delayed visibility of Jupiter's moons, he calculated a value for c of the correct order of magnitude: 212,000 kilometers per second.

train itself. Einstein understood this and built a consistent theory by focusing on the fundamental meaning of *c*. As a crowning achievement, it contained Newton's previous laws of motion as a limiting case for low speeds. In a certain sense, this theory can be seen as the unification of dynamics and optics – two fields of physics that had been separated by incomprehension throughout the 19th Century.

NO MORE THAN LOGIC

Let us consider the following situation from different perspectives: a train runs parallel to a wall that, while capable of reflecting light, stands at a distance *d* from the platform. If a flash of light is sent out by the train, the ray of light will be reflected by the wall and hit the train some time later (see figure).

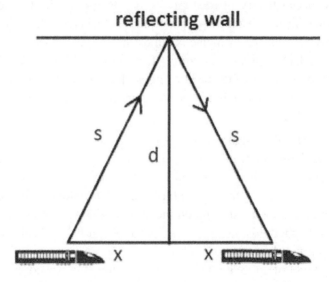

Schematic drawing of a light signal that is emitted by a train, then reflected by a mirror. Finally, it reaches the train again. The speed of the train is highly exaggerated.

Chapter 4: A precursor of deep thoughts

As the train has covered a distance of *2x*, the flash of light has had to travel a distance of *2s*: a little more than twice the distance to the wall (*2d*). So much is self-evident. Now let us switch from the perspective of the platform to that of the train driver: he or she can claim that the train is stationary, and the reflecting wall is moving at a specific distance *d* parallel to the tracks, with the same speed *v* that an observer would have assigned to the train. The two perspectives are physically equivalent! But from the driver's point of view the light travels a shorter distance: only *2d*. There is nothing unusual about that either, but the crucial point of the thought experiment is Einstein's insight that the speed of light is always the same – regardless of who is observing it. The observer on the platform will insist that the light path (*2s*) is shorter for the moving train (*2d*), and if *c* (distance per unit time) is the same, then the time shown on the train must also be shorter. So the moving clock on the train (from the viewpoint of the platform) runs slower!

It is amazing how much discussion is generated by this result of simple logic. "Moving clocks run slower", interpreted in the right way, is a direct consequence of the fact that the speed of light does not depend on the motion of the observer. And of course, it has been confirmed by countless experiments. The famous formula that relates the times t′ and t on the clocks on the train and the platform, respectively, is $\frac{t'}{t} = \sqrt{1 - \frac{v^2}{c^2}}$. It can be easily deduced by Pythagoras' theorem, $s^2 = d^2 + x^2$, taking *s=ct, d=ct′* and *x=vt* for the paths taken.

$$\text{Derivation: } \frac{t'}{t} = \frac{d/c}{s/c} = \frac{d}{s} = \frac{\sqrt{s^2 - x^2}}{s} =$$
$$\sqrt{1 - \frac{x^2}{s^2}} = \sqrt{1 - \frac{v^2 t^2}{c^2 t^2}} = \sqrt{1 - \frac{v^2}{c^2}}$$

The spectacular prediction that moving clocks run slower thus follows from elementary mathematics, if we accept Einstein's premise that *c* is constant in different frames of reference.

Einstein's Lost Key

MASS IS CONGEALED ENERGY

The curious reader can use Einstein's formula $\sqrt{1 - \frac{v^2}{c^2}}$ to calculate the factor by which the time measured by a moving clock passes more slowly. If we set the velocity v to 60 percent of the speed of light c, the result is $\sqrt{1 - 0.6^2} = 0.8$, which means that an hour would last only 48 minutes. In an aircraft flying at $v = 300$ m/s instead, the effect of time dilation is nearly negligible: the hour would be only 1.8 milliseconds shorter.

The same formula yields the factor by which an object appears to shorten if you pass it at high speed (length contraction). And finally the formula tells you by what factor the inertial mass increases as an object approaches the speed of light. If we recall the formula for kinetic energy, $E = \frac{1}{2} mv^2$, it is immediately clear that the 'rest mass' (called m_0) must grow with increasing speed. Whatever amount of energy is supplied to accelerate a body, its velocity v can never exceed c. An increase in energy therefore has to be 'stored' in a larger mass m. After a few steps of arithmetic, the equation $\frac{m_0}{m} = \sqrt{1 - \frac{v^2}{c^2}}$ leads us to the best-known formula in physics: the relationship between energy and mass, $E = mc^2$.

According to a well-known mathematical approximation, the so-called Taylor series, $\sqrt{1 - x} \approx 1 - \frac{1}{2}x$, and also $1 + \frac{1}{2}x \approx \frac{1}{\sqrt{1-x}}$, if x is small. If the rest energy E_0 is defined as $E_0 = m_0 c^2$ (where m_0 is energy at rest), then it follows that:

$$E_{tot} = E_0 + E_{kin} = m_0 c^2 + \frac{1}{2} m_0 v^2 = m_0 \left(c^2 + \frac{1}{2} v^2 \right) =$$

$$m_0 c^2 \left(1 + \frac{1}{2} \frac{v^2}{c^2} \right) \approx m_0 c^2 \frac{1}{\sqrt{1 - \frac{v^2}{c^2}}} = mc^2$$

Chapter 4: A precursor of deep thoughts

The fact that every mass m_0 contains an enormous amount of energy $m_0 c^2$ is as spectacular as it is worrying – as was disastrously confirmed forty years later with the atomic bomb. In essence the special theory of relativity is included in both these formulae: $\frac{t\prime}{t} = \frac{l\prime}{l} = \frac{m_0}{m} = \sqrt{1 - \frac{v^2}{c^2}}$ and $E = mc^2$.[i] This is where Einstein had reached in 1905.

THE ACTUAL DISCOVERY

> *I was sitting in my chair in the Patent Office in Bern when I suddenly thought: a person in free fall is unable to feel his own weight. The truth dawned upon me. This simple thought impressed me for a long time. The enthusiasm that I felt then guided me to the theory of gravitation. – Albert Einstein*

A Chinese saying is: "Talent hits the target that everybody else misses. Genius hits the target that nobody else sees." It's fair to discuss whether the special theory of relativity was as exclusively the work of Einstein as it is usually presented – Poincaré and Lorentz also made important contributions, in terms of mathematics probably as important as Einstein's. But the subtlety that led to the theory of general relativity, known as the equivalence principle, is Einstein's sole insight. What does it consist of?

Einstein conducted one of his legendary thought experiments. He imagined an observer in an isolated room. Can he determine whether he is located in a rocket under acceleration in an empty space, or in a gravitational field such as on Earth? No. In both cases his sensations are the same. As in the case of the special

---

[i] The so-called Lorentz transformations, which convert one coordinate system into another, look more complicated, but there is no difference in principle.

theory of relativity, Einstein now ventured the conclusion: if experiments cannot distinguish one thing from another, then they must be equivalent. The observer cannot distinguish whether the perceived acceleration is dynamic or due to gravity. The inertia of the astronaut presses him against the floor just as the force of gravity would do on Earth.

Visualization of the equivalence principle. In both cases, objects fall "to the ground", the falling objects following a curved path. On the right, this is usually perceived as an effect of a force; on the left, it is seen as consequence of an accelerated reference frame.

THE MYSTERIOUS DUAL NATURE OF MASS

This has profound consequences if we consider the fundamental characteristics of mass. Did mass mean resisting acceleration? Such a property could be measured even in the absence of gravitating bodies, and we might call it inertial mass. Or does mass first and foremost mean gravitational mass? This would be defined by its attracting influence on other masses, even in the absence of motion. Or were the two different phenomena per-

Chapter 4: A precursor of deep thoughts

haps logically connected with each other, by a mysterious link? Einstein answered his own question:[i]

> "Now I fully understood what the equivalence of inertial and gravitational mass meant, and I marveled at its existence ..."[18]

This thought experiment regarding inertial and gravitational mass marked the beginning of general relativity. Of course, even such an ingenious concept does not turn into a scientific revolution unless it is worked out and verified by calculations. As we learned in Chapter 2, however, an intuitive idea was always the starting point. Einstein hit on it in 1907, two years after he had formulated the vital importance of the speed of light for matter in his special theory of relativity: Inertial and gravitational mass were the same thing. Einstein now drew a consequence from a link between the special theory of relativity and gravitation: if clocks ran more slowly when in motion, then gravity, if indistinguishable from acceleration, should also have an impact on how fast time passes.

We could now go ahead and calculate the effects in gravitational fields with a similar formula (I may as well reveal it straight away: $\frac{t\prime}{t} = \sqrt{1 - \frac{2GM}{rc^2}}$), but doing so immediately would show little respect for Einstein's conceptual insight that took place long before the formalities were completed. In retrospect, it is the computational achievement that looks magnificent. But long before the formalisms had been developed, somebody had to get the ball rolling. Einstein was the only one who had noticed the problem gravity had posed. In turn, he substituted not

---

[i] In 1907 the Hungarian Baron Eötvös demonstrated the equivalence of inertial and gravitational mass with an ingenious experiment – of which, however, Einstein was unaware at the time.

only Newton's law of motion (with special relativity) but also his law of gravitation.

VIENNA'S DEEP THINKER

Yet there is a researcher whose conceptual discoveries were just as important as Einstein's: Viennese physicist and philosopher Ernst Mach (1838–1916). His status in the present-day hierarchy of theoretical physicists is much lower, mostly because Mach used to express his ideas in a nonmathematical way (although he could do little else at the time). Earlier than Einstein, Mach had intuitively recognized that the laws of dynamics could depend only on the motion of masses relative to each other. In contrast to Newton, he was convinced that the concept of absolute space did not make sense – because it was unobservable as a matter of principle. Several years before Einstein, Mach had generalized Galileo's discovery – that the laws of nature are independent of motion – to accelerated motion. This was a powerful act of abstraction that challenged Newton's authority. In 1883, Mach reduced Newton's "absolute" space to absurdity, more clearly than Einstein did.

Newton had imagined a rotating bucket of water, in which due to centrifugal forces, the water level rose up near the walls of the bucket. He concluded that the curved shape of the water level was evidence of an absolute space, the unaccelerated state being characterized by a flat water surface. Mach's analysis however went further, commenting upon Newton's thought experiment with dry logic:

> "Newton's experiment with the rotating water vessel simply tells us that the relative motion of the water with respect to the sides of the vessel creates no noticeable centrifugal forces, but that such forces are produced by the relative rotation with respect to the Earth and the other celestial objects. No-one can say how the experiment

Chapter 4: A precursor of deep thoughts

> *would turn out be if the walls of the vessel increased in thickness and mass, eventually reaching a thickness of several miles. We only have the one experiment, and we have to reconcile it with the other facts known to us – but not with our arbitrary fictions."*

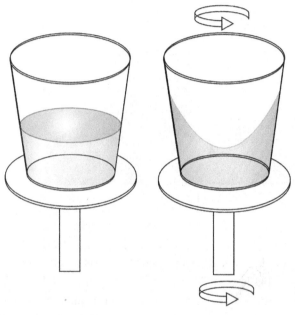

Newton's famous thought experiment: a stationary bucket (left) and a rotating bucket (right) in which the water level drops at the center and rises up near the walls.

Obviously, he considered absolute space as one of these 'fictions'. Mach had already dealt Newtonian physics a heavy blow. Einstein was aware of this, and he appreciated Mach's thoughts. In 1916 – too late, we might almost say – he wrote:[19] "The classic laws of motion suffer from an epistemological shortcoming, which was for the first time pointed out by E. Mach in a clear manner."

Einstein's Lost Key

Ernst Mach (1838–1916)

However, even theories with serious conceptual flaws often survive for a long time, as the history of science clearly shows. The revolution does not take place until there is a better alternative in sight. This is what Mach's criticism did not deliver, and he did not see the solution, the equivalence principle. Einstein was familiar with Mach's work, of course, and throughout his life he emphasized how Mach had stimulated his own ideas.

But Mach was much more than just a mathematically sluggish precursor of Einstein's ideas. His ideas about gravitation were visionary. One must emphasize the underestimated role of such conceptual ideas in the history of science, which – like raw diamonds – are easy to overlook until they are cut and polished to mathematical brilliancy. And unfortunately, it is often historical coincidence that determines which stones are chosen and which are left lying in the mud.

Chapter 4: A precursor of deep thoughts

RELATIVITY IS NOT THE SAME AS RELATIVITY

After his argument with Newton's bucket[20], Mach's thoughts went one step beyond: if a rotating bucket in a universe at rest were indistinguishable from a bucket at rest around which the universe rotates, then distant celestial objects would obviously influence the inertial characteristics of masses. Mach now conjectured that distant celestial objects must also be responsible for masses having gravitational properties. Hereby he had anticipated Einstein's later comparison of inertial and gravitational mass known as the equivalence principle.

Mach's hypothesis contains profound logic, and establishes a tantalizing relation between everyday objects and the universe. The acceleration of a falling apple would not only be determined by Newton's (supposedly) universal gravitational constant, but also by the distribution of masses throughout the universe! Mach's style was all but mathematical and unfortunately, due to an as-yet non-existent cosmology, he could not even think about formulating the idea in quantitative terms. The so-called Mach's principle therefore has a miserable reputation as a philosophical curiosity of the 19th Century. However, it contains a revolutionary idea that could have changed physics even more than the theory of relativity did. Einstein was very close to that in 1911.

We need to digress here a little in order to appreciate how sensational Mach's and Einstein's discovery really was. It has become so ingrained in our thinking that we tend to overlook it. Mass is a concept that can be defined in purely kinematic terms. Even in zero-g space, a shot-putter would feel that the greater the mass of the shot, the harder it is to accelerate: the more inert it is, in other words.

The term 'mass' makes sense on its own: it can be defined without resorting to gravity. But it is utterly surprising that the same concept, mass, is also responsible for a type of force, the gravitational force. In contrast, for example, electrical charges have nothing to do with inertia,[i] while mass is 'heavy' and 'inert' at the same time. Mach realized this intuitively and embedded it into his thought construct. But it was Einstein who in 1907 identified the problem precisely, formulating the equivalence principle, the beginning of general relativity.

Mach's principle thus has two different aspects. First and qualitatively, just as the equivalence principle, it says that inertial and gravitational mass are mysteriously connected. But secondly, Mach also claimed that inertia (i.e. the resistance to acceleration) must have its origin in the relative acceleration with respect to all other masses in the universe. This meant that the *strength* of gravity was also determined by every other celestial body – and suddenly we have a quantitative statement. This second aspect, however, did not enter into Einstein's subsequent formulation of general relativity. We might say that Einstein brilliantly formulated and solved one of the two problems that Mach had raised.

General relativity, however, is unable to explain the strength of gravity, as expressed in the numerical value of the gravitational constant. This is why Einstein's idea of 1911 is the true diamond among ideas about gravitation – because it would also have paved the way to a solution to the second problem.

THE LATE REVENGE OF NEWTON'S ABSOLUTE SPACE

With respect to one modern observation, Mach's viewpoint actually looks better than Einstein's. The cosmic microwave background (CMB) is a signal associated with the early period

---

[i] In this context we shall mention radiation damping of accelerated charges, a very interesting yet poorly understood phenomenon.

Chapter 4: A precursor of deep thoughts

of the cosmos. It is assumed that this signal is a cooled form of the radiation of what was then a hot, dense, and homogenous universe.[i]

However, the CMB is not uniform in every direction. A small red or blue shift of the light along a celestial axis suggests that our galaxy is moving with respect to the frame defined by the expansion[ii] of the universe. To put it precisely, we are heading for the Crater constellation at a speed of 370 kilometers per second. What is so special about that? Well, first of all, by cosmic standards this speed is modest – approximately one thousandth of the speed of light. Secondly, and far more irritatingly, this is the first observation in physics since the time of Galileo that *does* depend on the motion of the observer!

So has Newton's absolute space shown up after all? According to Einstein, the reference frame of the cosmic microwave background must be irrelevant – there is no place for it in his equations. But from Mach's point of view, it can well have significance. The laws of dynamics do depend on relative motion only, but this does not preclude the rest frame of all other mass in the universe producing a subtle effect. Admittedly however, there is no worked-out theory about that.

I am presenting these arguments with a certain level of detail that may appear to focus too much on the past. But present-day big science, with its numerous specializations, has set the formal mathematical version of general relativity so firmly in stone that the vast majority of gravitational physicists who deem

---

[i] I now seriously doubt the validity of this interpretation, in particular owing to thorough analyses by Pierre-Marie Robitaille, a radiologist from the University of Ohio. Unfortunately, Robitaille's criticism, though sound, had not been answered properly. It is difficult to decide who is right, but on the grounds of general methodology I refuse to accept results from researchers who avoid transparency and discussion.

[ii] The consequences of Einstein's idea shall later challenge this interpretation.

themselves experts do not even dare to think that Einstein stood at a real crossroads in 1911. Dozens of theoretical constructs have been piled on top of each other on the 1915 theory. Thus, the sheer fear of making the whole edifice crumble hinders a deeper reflection about the fact that the foundation may be incomplete. But I am afraid that it is.

It sounds like a bold claim, but the two problems that Mach identified are simply more important than everything that has been written about gravitational physics since 1915. Nothing can be more elementary than the discovery that gravitation has its origin in the properties of the universe.

EINSTEIN AND MACH VS. NEWTON?

Let us contemplate the ideas of Newton, Mach and Einstein about space once again. Mach and Einstein agreed that Newton's static absolute space was not observable as such. No meaningful theory could be built on it. Mach called his opposing point of view relativity, by which he meant that the laws of nature must depend on the relative motion of masses only. Einstein's special theory of relativity, in contrast, gets its name from the fact that the same laws of nature hold true in reference frames that move relative to each other. Einstein was referring to mathematical coordinate systems, Mach to physical masses.

Einstein then generalized this independence of *motion*, turning it into independence of the *acceleration* of the reference frame. The idea behind this is the equivalence principle, according to which an astronaut in free fall in a gravitational field feels exactly the same as one who is floating in zero-g space. Mach's generalization, in contrast, stressed that the crucial factor was acceleration *relative* to other masses.

Since 1915, Einstein's view that the motion and acceleration of the reference frame are irrelevant dominates, backed by the observation that the motion of the Earth and the Sun do not generate visible effects. According to Mach, however, this

Chapter 4: A precursor of deep thoughts

could simply result from the tiny mass of the solar system. It would be still possible that motion relative to (the barycenter of) all masses in the universe causes measurable effects.

Newton, Einstein, Mach. Caricature by Laurent Taudin, from "The Road to Relativity: The History and Meaning of Einstein's "The Foundation of General Relativity". (Princeton University Press, 2015, S. 150). Permission granted by the authors Hanoch Gutfreund and Jürgen Renn.

In Chapter 7 we shall discuss how Einstein's 1911 idea realizes Mach's principle, and in Chapter 12 we shall see that cosmological observations clearly support it – if correctly interpreted. Although Mach's critique of Newton had a decisive influence on Einstein, Mach's visionary ideas did not enter the theory of general relativity of 1915. The discrepancy that remained between Einstein and Mach could, had history taken another path around 1911, have become a spectacular unification.

Einstein's Lost Key

Chapter 5

The Revolutionary Idea of 1911

Why light propagates along curved paths

This chapter describes how Einstein developed his most important idea, how this idea was received at the time, and how it is seen today. What was going on in Einstein's mind back then? He started from the equivalence principle he had formulated in a thought experiment in 1907. In addition to the fascinating equality of 'heavy' and 'inert' mass, it contains another feature that becomes important when we look at the motion of bodies. Once again, Einstein considered a rocket under acceleration in weightless space and compared it to a gravitational field in a chamber on Earth. In both cases, as we have seen in the last chapter, 'falling' objects follow a curved path.

The same applies to a light ray entering a room from the side. It propagates straight in its rest frame, but with respect to the accelerated observer it is 'falling', eventually leaving the room at a lower level. Obviously, the light path in the accelerated rocket is perceived as curved. If this situation is really physically equivalent to a gravitational field, then light rays entering the field at a right angle must bend in its direction. This is a pivotal consequence of the equivalence principle.

It should be however emphasized that this result was derived directly from the properties of light, and not by a mere analogy stating that light might behave like a particle. As long ago as 1801, Georg Soldner, an astronomer who later served the Bavarian king, predicted that light would be deflected by the Sun.

Einstein's Lost Key

The main reason why that article became known was a polemical attack on Einstein by German nationalist and "Arian physics" propagandist Philipp Lenard. In order to discredit Einstein, Lenard had Soldner's article reprinted in 1921, accusing him of plagiarism. Not only this was absurd, but the accusation missed the point. A sound explanation of the deflection, taking due account of the different nature of light and matter, requires the type of argument made by the equivalence principle.

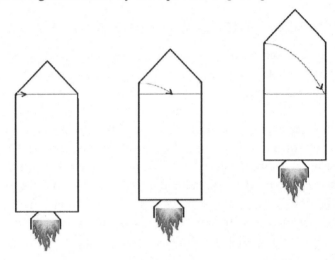

Illustration of the equivalence principle. Even in an accelerating rocket, light rays entering from the left are 'curved' (the image shows three sequential points in time). The same must be true in a gravitational field.

OPTICS REDISCOVERED

But how does a celestial body 'at rest' manage to bend light when it passes? Einstein's first idea was simple and intuitive. Without diving into mathematics, Einstein immediately remembered that curved light rays are a well-known phenomenon of optics, the law of refraction. Discovered by Dutch mathematician Snell as long ago as 1621, its deeper meaning was later outlined with the wave theory of Christiaan Huygens. Later

Chapter 5: The revolutionary idea of 1911

again, in 1660, the brilliant mathematician Pierre de Fermat derived the law of refraction with a strikingly simple principle: Light seeks the fastest path!

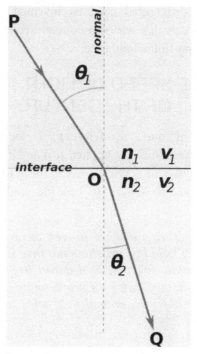

Law of refraction: light is deflected towards a medium that has a higher refractive index. The reason for this is the diminished speed of light.

It is obvious that (as in the example shown in the illustration) the shortest path, geometrically speaking, is not always the fastest, because it would partly take it through a medium at a lower speed (such as water). The 'deflected' ray of light finds quite a clever compromise in a path that is geometrically somewhat longer, but saves time, because a larger part of the movement takes place at higher speed.

While Fermat's principle is not difficult to understand, he first had to develop the mathematics needed for it. It is relatively easy to see that a sudden change in the speed of light creates a

Einstein's Lost Key

turning of a light ray, while curvature can be ascribed to a steady decrease in speed. But a quantitative formulation requires the so-called calculus of variations, which is a generalization of the differential calculus invented by Newton and Leibnitz. This calculus was subsequently refined by Leonhard Euler and Joseph-Louis Lagrange.

VARIABLE SPEED OF LIGHT – THE IDEA OF THE CENTURY

Einstein was of course also familiar with the Fermat-Huygens principle, and he intuitively applied it in a completely different context: the physics of gravitation. The speed of light in a gravitational field must be lower! His best idea had seen the light of day:

> *"Starting from the just proven theorem that the speed of light in a gravitational field is a function of position, it is easily deduced from Huygens' principle that light rays propagating at right angles to the gravitational field must undergo deflection."*[21]

In the preceding paragraphs of this 1911 article Einstein had discussed how clocks run in gravitational fields. In special relativity, he had worked out in detail how clocks run in different reference frames, and this obviously helped him now. If the speed of light was lower in a gravitational field, then clocks should run slower there, too.

To introduce a common notation, we consider the short formula $c = \lambda \cdot f$, which states that the speed of light c (in m/s) is equal to the product of the wavelength λ (lambda, in *m*) and frequency f (oscillations per second, i.e. *1/s*). To the reader unfamiliar with formal expressions, I suggest that there should be no fear of simple formulae: their concise language often gives a clearer understanding than long sentences ever could. An inval-

Chapter 5: The revolutionary idea of 1911

uable tool for dealing with formulae should be mentioned: physical units (meters and seconds in this case) that immediately enable one to recognize the correct combination of the respective quantities.

Albert Einstein (1879–1955)

IT WASN'T THAT SIMPLE: TIME AND LENGTH SCALES

In fact, Einstein had to take an important step beyond Fermat. In conventional optics only shorter wavelengths λ contribute to a decrease of the speed of light c: frequency f is not involved, as no-one would expect clocks to run slower under water. If a light wave changed its frequency on passing from air to water, for example, it could not maintain its wave form and would get out of phase. We shall come back to this important point in a cosmological context.[i]

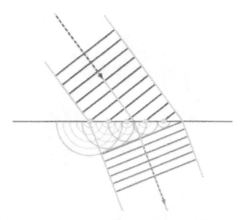

Passage of a light wave into a denser medium. The slower propagation speed at bottom forces the wave fronts to change direction.[ii]

Einstein, conversely, thought the decrease of the speed of light c in the gravitational field was caused by lower frequencies f – by clocks running slower, in other words. This was logical insofar as special relativity required clocks in moving reference frames to tick slower. According to the equivalence principle, reference frames under acceleration, that is, at ever-increasing speeds, have to be treated like a gravitational field, so clocks in there ought to be affected. By comparing the situation to special relativity, Einstein calculated the amount by which the clock frequencies f in a gravitational field decreased.

---

[i] It turns out that propagating light (that is traveling) and the light emitted by atoms must be distinguished.

[ii] A very good animation is http://en.wikipedia.org/wiki/Snell's_law#/media/File:Snells_law_wavefronts.gif.

Chapter 5: The revolutionary idea of 1911

All this is contained in his 1911 article, where he developed the formula that anticipated what is known as the 1915 theory of general relativity:[i]

$$\frac{c}{c_0} = 1 - \frac{GM}{rc^2}$$

> daher die Lichtgeschwindigkeit c in einem Orte vom Gravitationspotential Φ durch die Beziehung
>
> (3) $\qquad c = c_0 \left(1 + \frac{\Phi}{c^2}\right)$
>
> gegeben sein. Das Prinzip von der Konstanz der Licht-

Einstein's original publication of 1911, p. 906. The gravitational potential $\frac{GM}{r}$ (with a negative sign, gravitational constant G, mass of the sun M, distance from the Sun r) is written as Φ.

The formula says that the speed of light near the sun c differs from the 'normal' speed c_0 only minutely, by a factor a little smaller than 1 that contains both the gravitational potential and the speed of light. This tiny difference causes an equally tiny degree of light deflection by the sun. It makes little difference here whether we say that the speed of light near the Sun c is a little smaller than it is at large distances (c_0), or if we take the speed of light near the Sun as the reference value c_0 and say the speed of light c is faster at large distances. Einstein could thus just as well have written $\frac{c}{c_0} = 1 + \frac{GM}{rc^2}$ in the latter case, though the formula he used shows a closer analogy to special relativity. In general, $1 + x \approx \frac{1}{1-x}$ is a valid approximation if x is a small

---

[i] However, as early as 1907, he had written in another context about the "effect of the gravitational field" on clocks, giving this very formula. Ideas that seem easy to understand in retrospect can take a long time to mature. The article, incidentally, was published by Johannes Stark, later an adherent of "Aryan" physics who attacked Einstein in a disgusting manner.

number compared to 1. Since $\frac{GM}{rc^2}$ is a small number, the two options are certainly hard to distinguish by observation. We shall come back to this subtle point in Chapter 8.

THE ROYAL ROAD TO UNDERSTANDING: PHYSICAL UNITS

The above formula again invites one to consider physical units, a very intuitive way of understanding connections between the laws of nature. The so-called gravitational potential $\frac{GM}{r}$ simply yields the energy of a body with mass m at a given distance r from a larger mass M, in this case of the Sun. G is the Newtonian gravitational constant $G=6.673 \cdot 10^{-11}\frac{m^3}{s^2\,kg}$, that mysterious number whose secret Einstein was so close to unveiling back then. The following quotation is a clear hint: [i]

> "The principle of the constancy of the speed of light only can be maintained only by restricting to space-time regions with a constant gravitational potential."

Maybe you can already see this secret shining through when considering the units: The gravitational potential $\frac{GM}{r}$ is divided by the square of the speed of light c^2, resulting in a pure, dimensionless number! Obviously, the gravitational potential must have the same unit as c^2 (m^2/s^2). Could not quantities with the same units just be the same? But Einstein could not arrive at this conclusion at the time.

Let us briefly summarize the key points of Einstein's discovery from an epistemological perspective, as elucidated by the following sequence: principle, phenomenon, calculability, cal-

---

[i] Annalen der Physik 38 (1912), p. 355–369.

Chapter 5: The revolutionary idea of 1911

culation. Although Ernst Mach had already understood essential aspects of gravity, it was left to Einstein to formulate the equivalence principle in a thought experiment. Not only did he conclude that light is deflected by celestial objects (concrete phenomenology), he also presented a mechanism for it: the variable speed of light (calculability). Finally the calculation itself was a mathematical achievement that produced the result of 0.83 arc seconds of deflection by the sun. This was the crucial step towards the general theory of relativity, which is often dated to 1915.

As the derivation is not trivial, the formula for the light deflection angle $\Delta\varphi$ is provided without calculation:

$$\Delta\varphi = \frac{2GM}{rc^2}$$

Unfortunately, this calculation in 1911 was based on a small omission in the underlying assumptions that had grave consequences. Einstein had realized that as clocks run slower in gravitational fields, timescales are longer and frequencies smaller. But clocks are ultimately no different to atoms that emit light of a specific frequency f *and* wavelength λ. Thus it is only natural for atoms to decrease their wavelength λ, too – a possibility that Einstein had overlooked. Since observations that could have guided him did not exist at the time, one can hardly blame Einstein for this merely technical omission. It would certainly have been noticed if the idea had achieved widespread recognition, but unfortunately it was precisely that small error that hindered the breakthrough.

EINSTEIN AT THE CROSSROADS

Einstein's discovery was highly exciting, and marked a promising milestone in the development of general relativity. He completed the paper in Prague in June 1911 and sent it to the

Einstein's Lost Key

Annalen der Physik,[i] where it was published immediately. Did the scientific community react with euphoria? Not exactly: perhaps Einstein's visionary achievement was formulated a little technically. Be that as it may, its reception was sobering. Only one German physicist who was well known at the time, Max Abraham, described Einstein's approach of variable speed of light as a "lucky idea" – while at the same time criticizing Einstein's equivalence principle of inertial and gravitational mass. Abraham claimed that Einstein had abandoned his most important principle: the constancy of the speed of light. Abraham's own approach was rejected in turn by Einstein, but is of some historical interest. Abraham's criticism however is based on a misunderstanding that continues to this day.

The ostensible contradiction between 'constant' and 'variable' speed of light is a typical confusion that arises from a lack of differentiation in everyday language. According to Einstein's special theory of relativity, it did not matter whether the speed of the train's headlight was measured from the platform or from the moving train: it always had the same value. The speed of light can therefore be called independent of the reference frame, i.e. as 'constant' with respect to the reference frame. But this does not mean that it cannot depend on position!

Dependence on position and dependence on the reference frame are two very different concepts, and there is absolutely nothing to prevent one being true and the other not. Certainly, the technical apparatus that described the independence of the reference frame, called Lorentz transformations, had not yet been worked out by Einstein with regard to his new idea. Why should he? First, one had to understand the basic mechanisms, and then address a whole series of problems later on. In a 1912 article, Einstein in fact acknowledged that he had not yet found the generalized Lorentz transformations, and that this was not as

---

[i] Not to be confused with the *Annals of Physics*.

Chapter 5: The revolutionary idea of 1911

easy as he had first thought. At the same time however, he left no doubt that he considered Abraham's criticism unfounded: "It must not put us off from pursuing the path we have taken." [i]

PRESCRIBED BLINDNESS

The above misunderstanding benefits from another convention of modern physics. The speed of light in today's unit system is set to the constant value of *299,792,458 m/s* by definition, that is, new measurements cannot change it. This has purely practical reasons for the system of units: in combination with the definition of a second[22] (9,192,631,770 oscillations of a cesium atom in a specific condition), the meter is now defined as the distance covered by light in *1/299,792,458* of a second.

With this definition, a variation of c cannot be determined directly: the "meter" simply changes in line with it! Nor is it noticeable if atoms oscillate more slowly: the length of the second, defined by the frequencies f of the atoms, then changes accordingly. As $c = \lambda \cdot f$ always holds, the variability of c is hidden in the definition of the units, as Einstein imagined. Whether this definition makes sense remains to be seen. Unfortunately, however, it is frequently used as an argument against variable speed of light, which is nonsense. The variability of c can, of course, be detected in an overall view, i.e. it is measurable: it directly manifests itself in curved light rays.

The variable speed of light is however hidden at first sight, and this can be illustrated by an analogy with temperature. A classic thermometer measures different temperatures because the alcohol expands by more than the glass surrounding it. But what if glass widened to the same degree as alcohol? No change in temperature would be displayed. You could put a thermometer in places of very different heat and still claim that the tem-

---

[i] Annalen der Physik 38 (1912), 355.

perature was the same everywhere. All rulers defining the length of the meter would also expand in line with the temperature. But curious curved spaces would be visible in this world: spherical areas of low temperature would bundle light like drops of water. In such a world, all physical laws could well be formulated with constant temperature. The only question is whether this is really practical for things like a weather forecast. If we return to the realm of gravitational physics, the assumption of a constant speed of light is just such an awkward convention.

THE ARROGANCE OF THE HERE AND NOW

Variable speed of light instead is a well-known and easily understandable concept in physics: in optics it has been successfully applied all along. The arguments discussed one hundred years ago by Einstein and Abraham should be well known to anyone who has dealt with variable speed of light. Yet today there are physicists who unknowingly rehash Abraham's critique, insisting that the very idea of variable speed of light is wrong – and without knowing anything of what Einstein thought of it.[23] The message of these clever articles seems to be that Einstein had no idea about the fundamentals of his own theory.

But such papers are by no means the exception, unfortunately: they are just one of many examples of the ahistorical, superficial way in which many physicists ply their trade today. They consider themselves brighter than Einstein only because they perform nimble-fingered arithmetic that was served up to them as students, thanks to the pioneering work of Fermat, Riemann and others. The idea that true progress in physics requires many years of intense reflection is alien to present-day paradigm of orgiastic calculation.

The alienation of physics from the ideas of natural philosophy that until 1930 were represented by Mach, Einstein, Schröding-

Chapter 5: The revolutionary idea of 1911

er, Bohr and Dirac merits a discussion on its own. Genuine science succumbed to a crisis. The conflicting interpretations of quantum mechanics eventually led to a paradigm of descriptive rules without fundamental reflections, and technology-oriented physics ultimately came to dominate with the atomic bomb in 1945. Physicists left their ivory towers in search of power and wealth.

Einstein was a solitary researcher for the last thirty years of his life. Because of the aforementioned paradigm change, his entire work is nowadays evaluated from the perspective of the generation of physicists that followed. They had no choice but to acknowledge Einstein's achievements – but are not in the least interested in his convictions about natural philosophy. One might begin to understand how it came about that variable speed of light is still so unknown today. The way Einstein's publications were judged by early biographers is telling in the extreme.

THE BIOGRAPHERS' GLASSES

Abraham Pais, himself a theoretical physicist and a companion of Einstein's, is considered a leading expert on Einstein. Pais, who lost family members in the concentration camps and barely escaped the Nazis himself, met Einstein in Princeton in 1947. Einstein liked his congenial directness, and their common Jewish origin undoubtedly acted as a bond between them. Pais's biography is a pleasure to read, and it bears the label of their friendship.

But this should not hide the fact that Pais and Einstein held very different views about the nature of physics. Pais belonged to the generation of particle physicists that described accelerator experiments with a plethora of new particles, an approach that was diametrically opposed to Einstein's philosophical quest for understanding. That being said, Pais's biography should be read with caution where it presumes to evaluate Einstein's ideas.

Einstein's Lost Key

For example, Pais calls Einstein's publications on variable speed of light around 1911 a "work in progress". With the alleged wisdom of hindsight, he searches for early imprints of the formalism that was developed later. Pais claimed that the content of the 1911 paper dealt with "more of the same": the equivalence principle, gravitational energy, red shift, light deflection. He completely blanked out the most important content – variable speed of light – even though it appeared in the title.[i] On Einstein's crucial formula (see above) Pais remarked: "Einstein restored sanity, but at a price" – as if the variable speed of light were a blemish rather than a revolution. Here is how Pais commented on the conceptual sensation that Einstein was tracking, namely the unification of the electromagnetic phenomenon of light with gravitation:

> *"What possessed Einstein? Why would he ever write about a static gravitational field coupled to a non-static[ii] Maxwell field and hope to make any sense?"*

Well, one can just say what possessed Pais: he did not understand Einstein.[iii] The great picture that Einstein had set out for – unifying the fundamental interactions – had completely escaped Pais. Instead he put his own oar in, and in the guise of a fictitious Einstein talking in first person, Pais pontificates for several pages about what *he thinks* Einstein *would later have said* about his idea. That it was naive, that Einstein had not recognized this and that, etc. Finally Pais pretentiously sums up Einstein's work in Prague as follows:[24] "As yet, he had no theory of

---

[i] "The effect of gravity on the propagation of light", Annalen der Physik 35 (1911), p. 898; "The speed of light and the statics of the gravitational field", Annalen der Physik 38 (1912), p. 355.
[ii] A formal and insignificant aspect.
[iii] Pais also repeated Max Abraham's mistake: "Einstein tried the impossible – to incorporate this nonconstant light velocity into the special theory of relativity." (loc. 6767).

Chapter 5: The revolutionary idea of 1911

gravitation. But he had learned a lot of physics" Essentially this is the view that prevails to this day among theoretical physicists.

THE LOST JEWEL

Pais was not, of course, the only one to blame. The common practice of historiography, namely the retrospective construction of a narrative, has greatly contributed to erasing Einstein's brilliant idea from the collective memory. Another well-known biography by Walter Isaacson briefly discusses the 1911 article and the first prediction of light deflection therein. But the underlying idea of variable speed of light does not get a single mention! In the later literature the idea practically disappeared. Notwithstanding the fact that Einstein continued to mention the concept of the variable speed of light. For instance, in an article[25] of 1916 - i.e. after the geometrical formulation and even later, in an essay[26] in Nature in 1920. One can therefore even argue that Einstein never really gave up the idea.

Let us return to Einstein's work back then. What happened next? That article, sent from Prague in June 1911, was an interim climax of his idea. In the following month, Einstein attended a conference in Karlsruhe, and then was busy with completely different topics: he had promised to speak at the first Solvay Congress in Belgium in late October 1911, and also to write an article[27] entitled *The present state of the problem of specific heats*.

In 1912 he authored two more articles on variable speed of light, though the second one was a response to the aforementioned critique by Max Abraham. In general, Einstein expressed himself very carefully about the idea, almost exhibiting too much self-doubt.[28] He let himself be drawn into debates that had no direct connection with the exciting aspect of the idea: the link to Mach's principle and the groundbreaking possibility of eliminating the gravitational constant. This most important merit of the new theory went practically unmentioned, the main focus lying on the problems of its technical implementation. In a rather unknown article in 1912, Einstein writes about Mach's thought in a highly interesting way: [29] "The assumption is obvious that the whole inertia of a mass point is an effect of the presence of all other masses, based on a kind of interaction with the latter." However, he does not establish the reference to the calculation of G, also due to the lack of cosmological data, of course.

The house where Einstein lived in Prague: Třebízského uliza 1215 on the left bank of the Vltava, today Lesnicka 7. Photo: the author.

At the same time, in 1912, Einstein was thinking of moving back to Zurich again. The professorship in Prague had been a step up in both financial and academic terms, but he – and his wife Mileva Maric even more so – did not feel at ease there. She blamed her poor health on the bad air and dirty water.[i] Although he had also been offered a position by the University of

---

[i] At a conference in Prague in July 2015, I took the opportunity of visiting Einstein's house. It is located very close to the Vltava, surrounded by stunning Art Nouveau facades. So this can hardly have been the reason why the Einsteins disliked living in Prague.

Chapter 5: The revolutionary idea of 1911

Utrecht, he decided to return to his former workplace, the ETH Zurich, now as a full professor. For the variable speed of light, however, this was to be an unfortunate move.

Although Einstein discussed the variable speed of light with Hendrik Antoon Lorentz and Paul Ehrenfest in Leidenin 1912 (the details are unknown, alas), the reaction to the articles in 1911 and 1912 showed that Einstein's best idea was not understood by his contemporaries – let alone appreciated. Many found its revolutionary content irritating, and it became the subject of petty debates. Einstein was working on the idea all alone at the time, and several distractions prevented him from pursuing it in earnest. Its obvious relevance to the cosmos, the size of which was as yet totally unknown, had not yet been noticed. The idea was in danger of dying long before unfolding its potential. With Einstein's return to Zurich, fate took its course.

Einstein's Lost Key

Chapter 6
Seduced by Mathematical Beauty

The Pyrrhic victory of geometrical formulation

For this chapter in particular I had to fight with my editor, who had heard about the following anecdote: when English astrophysicist Sir Arthur Eddington was asked in 1919 whether it was true that only three people in the world understood the theory of general relativity, he allegedly replied: "Who's the third?" Since then, general relativity has been considered prohibitively difficult in terms of math, a holy grail of abstract geometrical concepts inaccessible to ordinary mortals. How can it be put across in a popular science book?

It can be done. Naturally we shall not approach the subject in a conventional mathematical way, and this chapter cannot, of course, replace detailed work on basic principles as part of a course of study. But it does not take a mathematical genius to gain a clear understanding of the concepts that are central to general relativity.

Our memory is largely dominated by visual content. The ability to cover A4 pages with formulae didn't give Neanderthal man any evolutionary advantage, and thus Homo sapiens is not particularly good at it either. Our brain is therefore sluggish when it comes to pure calculation, but we can use its huge capacity in image processing. It is possible to visualize difficult concepts of differential geometry whose formal presentation – and especially proof – looks scary. To help those who are interested to get started, I shall mention the mathematical symbols of the respective quantities. While this is not necessary in order to understand the following chapters, it certainly helps to see why Einstein was intrigued by these geometric objects.

THERE IS NO KNOWLEDGE YOU CANNOT EXPLAIN

> *"A theoretical science that is unaware of the fact that their important concepts are supposed to be expressed in concepts that are understandable for the educated are intended to become part of the general erudition – a theoretical science, I say, in which this is forgotten and the initiated continue to whisper things to each other that are at best understood by a small group of peers, will inevitably be cut off the rest of the cultural community; in the long run it will wither and ossify, no matter how vividly the esoteric claptrap within their coteries of experts may go on."* – Erwin Schrödinger

I am not driven by an unhealthy pedagogical desire to make you understand the content of general relativity. I rather think that society has a right to evaluate scientific theories, and that in the long run this can only succeed with the participation of a wider public. Because all too often in the history of science, closed circles of so-called experts, claiming a monopoly of understanding, have gone astray.

So no-one should put up with being told that a theory is so difficult and incomprehensible that evaluating it must be left to the specialists. This leads at best to academic absurdities, at worst to a scientific dead-end of historic dimensions.

But there is a psychological reflex: incomprehensible theories are often particularly popular thanks to their mystical aura. How else are we to explain why constructs such as 'cosmic inflation'

Chapter 6: Seduced by mathematical beauty

or string theory have become far better known than more straightforward concepts such as the ideal gas, for instance, could ever hope to be? People are inclined to believe in what they do not understand. That helps to preserve the world view, but is dangerous. Even scientists rely on the authorities in related fields, simply retelling theories that they themselves have not had the time to grapple with.

If you have resigned yourself to having only a little knowledge, a clear explanation of a complex matter can easily be met with mistrust and resistance: if it were that easy, I'd have thought of it myself. The attempt to explain something that is hard to understand will inevitably earn the author accusations of having misunderstood something himself.

Over and over again it is drummed into us that unfortunately, even the basics of the theory are accessible only to experts after years of training. They will tell you that an illustrative approach is counter-productive and doomed to fail. As an answer I can only tell you what Einstein thought of these experts: if you cannot explain something simply, you haven't fully understood it yourself.

NUMBERS AND MACHINES THAT PROCESS THEM

"I was lucky to come across books that didn't take logical rigor too seriously" – Albert Einstein

But now to the matter in hand. Einstein's general theory of relativity is concerned with what is known as differential geometry. This means that the geometric objects it deals with are not only planar and linear, but curved. Describing them requires some explanatory aids.

Einstein's Lost Key

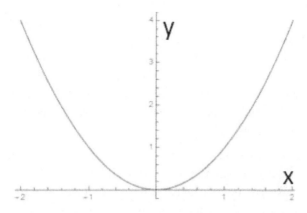

The normal parabola, with numbers on the X axis and their squares on the Y axis.

Simple mathematical functions are 'machines" that produce one number from another. The function $x \rightarrow x^2$, for example, assigns to any given number the square of that number: if you put 2 in, 4 comes out. We can also say *f(2)=4* or in general *f(x)=x²*. The results can be visualized with a graph (see parabola). Similarly, a function may depend on two variables, x and y, and the result can be displayed in a third dimension.

The concept of a function has a variety of applications. Imagine, for example, that a thermometer displays the air temperature, which varies with time. Each point in time *t* is then associated with a temperature *T(t)*: this is a function, too. It is easy to imagine observing the temperature in a space at a fixed time. A point in space is defined by three coordinates *(x,y,z)*. If one enters these three numbers into the "machine", the function *T(x,y,z)* will "spit out" the temperature at this location. This sort of function is difficult to represent graphically, though. *T(x,y,z)* is also called a 'field'.

Chapter 6: Seduced by mathematical beauty

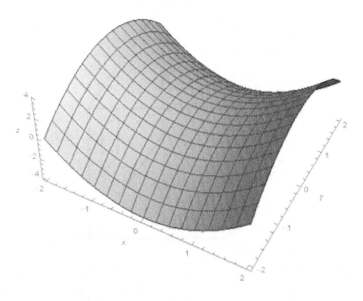

A representation of a function depending on two variables, x and y, as a surface in three-dimensional space. This function is $f(x,y) = x^2 - y^2$.

WHAT LEIBNIZ, NEWTON, AND CARTAN FOUND OUT

> *One of the saddest developments in the teaching of mathematics was the neglect of clarity in favor of formality.* - Ian Stewart

If we define every point in a room with coordinates (x,y,z), then the temperature may be $T(x=3, y=3, z=1) = 295$ Kelvin[i] (equivalent to $22°$ Celsius) in the center, for example, and $T(x=3, y=1, z=1) = 291$ K $(18°$ C) by the window, see figure below.

---

[i] The physical temperature unit, which defines the lowest reachable temperature of $-273.15°C$ as 0 K.

Einstein's Lost Key

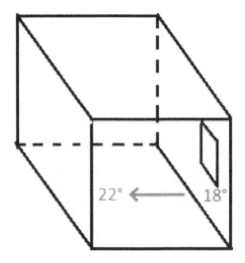

Here the temperature field is shown only at two points, 18° and 22°. Correspondingly, there is only one gradient arrow. In the above example it would amount to about one degree Kelvin per meter.

If such differences arise, then we are interested in a somewhat sophisticated notion, the so-called temperature gradient.[i] It shows by how many degrees Kelvin the temperature changes per meter. More accurately, it involves lots of little arrows, their direction indicating where it is getting warmer. They might point away from the cold window to the middle of the room, and the longer the arrow, the larger the difference per meter. Parents who tell their kids hunting for Easter eggs that they are "getting warmer" are communicating a temperature gradient.

Mathematicians call the process by which we determine gradients from the temperature field $T(x,y,z)$ "differentiation" and the result "derivative". In this process a variable giving a Kelvin reading (temperature) is converted into a variable whose unit is degrees Kelvin per meter (K/m).

---

[i] The term has nothing to do with degree.

Chapter 6: Seduced by mathematical beauty

The reverse case is just as important. Even if you only know the incremental temperature differences, the systematic composition of small steps enables you to determine the temperature difference between the start and end points. This calculation is called integration. It is the opposite of differentiation, because a variable with the unit of degrees Kelvin per meter is changed into a temperature difference (only degrees Kelvin). In both cases, differentiation and integration, small steps are assumed within which the temperature changes are negligible.

We do not need to go into the formal aspects, which were first developed by Newton and Leibniz. But a concept that was introduced much later by French mathematician Élie Cartan really is important. If you integrate the temperature gradient (by the systematic addition of small differences) along a path, then you can imagine a (one-dimensional) path being "eaten up" and a number (temperature difference in degrees Kelvin) being "delivered". Such quantities, which acquire meaning only when integrated along a one-dimensional path, are called one-forms. They play a central role in differential geometry. We shall come back to this soon.

ARROWS, ROTATIONS AND EVEN MORE SOPHISTICATED NUMBERS

An even better example of arrows (known as *vectors*) in three-dimensional space would be the wind speed, blowing at each location (x, y, z) in a certain direction and with a certain strength. Each arrow would again have three components (v_x, v_y, v_z), indicating the component of the wind velocity in the particular direction (see fig.). Mathematicians, who like abbreviations, often write velocity vectors as v_i, where i can take the values x, y and z.

Einstein's Lost Key

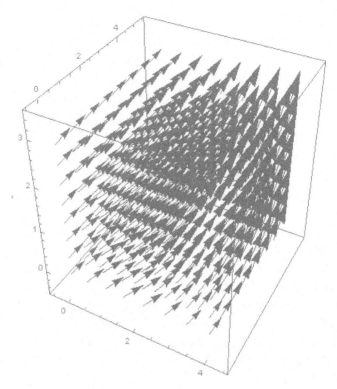

*Graphical representation of a vector field. One has to imagine an arrow with direction and length at **any** point in the space.*

Mathematics thrives on lots of generalizations. A vector consisting of three numbers can again be processed by a "machine" – a function, in other words – that turns it into a number or even another vector. So there are functions that affect arrows (vectors), and transform them into other vectors. But as a vector in three-dimensional space already contains three numbers (for the x, y and z components), a more complicated machine is needed in order to build one vector from another one. It's called a matrix, and consists of a *3x3* grid by which a vector is multiplied according to a particular rule, resulting in a new vector. Here are some calculation examples, if you're interested:

Chapter 6: Seduced by mathematical beauty

$$\begin{pmatrix} 3 & 1 & -1 \\ 2 & 0 & 1 \\ 4 & 2 & -1 \end{pmatrix} \cdot \begin{pmatrix} 1 \\ -2 \\ 5 \end{pmatrix} = \begin{pmatrix} 3 \cdot 1 - 1 \cdot 2 - 1 \cdot 5 \\ 2 \cdot 1 - 0 \cdot 2 + 1 \cdot 5 \\ 4 \cdot 1 - 2 \cdot 2 - 1 \cdot 5 \end{pmatrix} = \begin{pmatrix} -4 \\ 7 \\ -5 \end{pmatrix}$$

$$\begin{pmatrix} 0 & -1 & 0 \\ 1 & 0 & 0 \\ 0 & 0 & 1 \end{pmatrix} \cdot \begin{pmatrix} 2 \\ 0 \\ 0 \end{pmatrix} = \begin{pmatrix} 0 \cdot 2 - 1 \cdot 0 + 0 \cdot 0 \\ 1 \cdot 2 + 0 \cdot 0 + 0 \cdot 0 \\ 0 \cdot 2 + 0 \cdot 0 + 0 \cdot 0 \end{pmatrix} = \begin{pmatrix} 0 \\ 2 \\ 0 \end{pmatrix}$$

Example of a vector a = (1,-2,5) and a matrix that transforms it. The vector is "turned horizontally", each of its components is multiplied by the corresponding row, and the sums in each row form the new components of the transformed vector (see the numerical example). It should be noted that in general a vector can also change its length. The second matrix B is an element of the group of rotations SO(3). It is easy to see that B rotates the vector by 90 degrees around the z axis: the vector originally pointed in the x direction, and in the y direction afterwards.

One common subtype of this sort of matrix executes only rotations, without changing the length of the vector. Called "SO(3) rotation group", it is a matrix species of paramount importance that we shall be seeing more of.[i] A less fundamental aspect of it is perhaps familiar: whenever 3D computer animations show the scene from a different angle, SO(3) matrices are involved.

$$\begin{pmatrix} \cos\varphi & -\sin\varphi & 0 \\ \sin\varphi & \cos\varphi & 0 \\ 0 & 0 & 1 \end{pmatrix}$$

A matrix subtype that only carries out rotation, here by the angle φ around the z axis.

---

[i] Just as one-dimensional vectors are abbreviated as v_i, matrices are written with two indices a_{ij}, where i and j can each take on the values x,y,z (or 1,2,3).

Einstein's Lost Key

A WALK ON THE GLOBE

Thus armed, we can already get an impression of how Einstein described the curvature of space in general relativity. First picture a straightforward case: a hiker finds himself at the North Pole of the globe (the prototype of a curved "space", although we are only concerned with its two-dimensional surface), holding a spear – i.e. a vector – in his hand. His task is now to hike towards the Equator without changing the direction of the vector, for example along the prime meridian (of longitude, i.e. Greenwich). The only directions allowed are those parallel to the two-dimensional surface of the sphere, i.e. tangential directions pointing to the horizon.

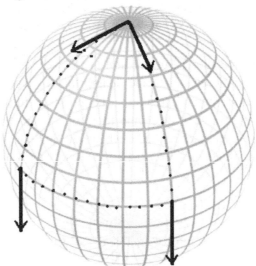

An example of the parallel shift of a vector: despite all efforts to maintain its direction it appears to have rotated after completing the North Pole – Equator – North Pole journey.

While trying not to alter its direction, at the Equator the vector points in the direction shown in the illustration, i.e. to the South. If we tell the traveler only to move sideways at the Equator, he has to keep the vector pointing south (at right angles to the di-

Chapter 6: Seduced by mathematical beauty

rection he is moving in) if he does not want to change its direction. Eventually, after passing some degrees of longitude – as far as the Maldives, say – he returns straight to the North Pole again. The vector now still points "South", but in a different direction (see fig.)! This paradoxical fact of the vector being rotated, despite all efforts to keep its direction constant, has its origin in the geometric properties of the sphere, or in curved spaces in general. The method obviously works even if the hiker is unaware of being on a sphere, thus it is suitable for measuring curvature in any space.

WHAT KEEPS THE WORLD IN ITS INMOST FOLDS?

The property of the sphere that it rotates vectors carried around on it is called a connection. It is easy to visualize, though describing it in words requires a certain amount of complexity. A rotation of the vector, as we have seen, requires a *3x3* grid (i.e. a rotation matrix), but the matrix is only created through a movement along a specific path.

One can imagine that a tiny rotation is generated at every point on the path, and adding these rotations together (i.e. integrating them) eventually delivers the result that we have seen on the return to the North Pole. However, this means that the connection doesn't just need two letters (indices) for its presentation, but also an additional one for the direction of the path. This is why the connection appears in textbooks as Γ_{ij}^{k}, which looks pretty awkward. The upper index k simply states that a line has to be "eaten", i.e. integrated, to produce a rotation matrix (with the lower indices i and j). The Γ_{ij}^{k} connection is thus a "one-form" (line eater), which spews out a rotation matrix as a result.

The closed line that we completed on our global hike also says something about the surface enclosed by it, in this case a spherical triangle containing Central Europe and part of Africa.

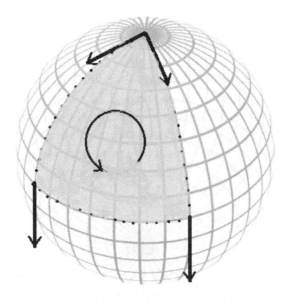

The spherical triangle is intrinsically curved, which leads to the rotation of a vector transported along the surrounding path.

It is one of the most beautiful mathematical theorems that the properties of a surface are in many cases contained in the properties of its edge, i.e. the line that marks its border. Nature evidently knows this! In electrodynamics, for example, the electrical current summed along a line is equal to the surface surrounded by this line, multiplied by the magnetic field passing through the surface. Known as Stokes' theorem, it undoubtedly excited Einstein to see the geometrical nature of this theorem reappearing in his beloved theory of gravitation. It was certainly the beauty and elegance of these structures that later led him to concentrate on these problems in his quest for a unified field theory[30]. He wanted to explain nature by means of geometry.

EINSTEIN REDISCOVERS GAUSS AND RIEMANN

Einstein was first gripped by enthusiasm for differential geometry in the fall of 1912, after returning to Zurich from Pra-

Chapter 6: Seduced by mathematical beauty

gue. His friend and classmate Marcel Grossmann, now Professor of Descriptive Geometry at the ETH Zurich, introduced Einstein to this branch of mathematics, in particular to the work of Riemann, Ricci and Levi-Civita.[i] Once Einstein realized that there was a connection with the theory of gravitation, he reacted immediately – even though he didn't feel up to the math: "Grossmann, if you don't help me I'll go mad!"

Bernhard Riemann (1826-1866)

---

[i] Georg Pick had already pointed this out to Einstein in Prague, but his work with Grossmann did not begin until after his return on August 10, 1912.

Einstein's Lost Key

Einstein and Grossmann were puzzled by that analogy with Stokes' theorem.[i] If the connection corresponded to electrical current, there had to be a geometrical counterpart to the magnetic field, too. Luckily Bernhard Riemann, a brilliant 19th Century mathematician, had already discovered it: the curvature tensor named after him. This is a "2-form": it "eats up" the surface enclosed by a path and delivers a rotation matrix (see fig.) as the result. The curvature tensor bears no fewer than four indices: $R_{ij}{}^{kl}$, where one has to imagine that two spatial directions k and l "span" the integrated surface (which means they define it), while i and j stand for the resulting rotation matrix. The Riemann curvature tensor is the central mathematical object of general relativity.

The curvature tensor is actually a four-dimensional die, which – if we add time as a dimension – has in total $4^4 = 256$ entries describing the curvature of space-time. Fortunately it has various symmetries, so many of them are identical: it is irrelevant, for example, whether a surface is spanned by x and y or by y and x.[ii]

THE EQUATIONS FOR WHICH EINSTEIN IS FAMOUS

Curiously, this mushrooming numerical construct boils down to much fewer numbers when dealing with physical variables. In a *3x3* matrix, for example, all nine figures are rarely important. The essential information is in the sum of the diagonal elements, which is called the 'trace'. Einstein generalized the system in which diagonal elements are added (called "contracting a tensor") and applied it to the Riemann curvature tensor. It works like this: the four individual elements of the curvature tensor are added and "contracted" to a new element (the two

---

[i] There is no historical evidence for this, but it is certainly plausible.
[ii] Further symmetries reduce the number of independent components to 21.

Chapter 6: Seduced by mathematical beauty

outside indices remain the same): $R_{11}{}^{12} + R_{12}{}^{22} + R_{13}{}^{32} + R_{14}{}^{42} = S_1{}^2$ and so on. The overall result in this case is called the "Ricci tensor" S_k^j. On how the tensor components had to be added in order to make physical sense, Einstein spent almost two years. It didn't come easily to him!

He probably wouldn't have reached his goal at all had he not intuitively guessed the outcome. Masses m and their energy $E=mc^2$ had to be responsible for the curvature of space, and in the case of motion the energy density had to be generalized into an energy flux density and momentum density[i], also a tensor.

Finally, by skillfully 'contracting' the tensor of curvature, Einstein found an object[ii] with properties that perfectly matched the energy-momentum tensor T_{ij}. Their identification has been referred to generally as Einstein's equations ever since. They are usually abbreviated as:

$$G_{ij} = \frac{8\pi G}{c^4} T_{ij}$$

Each of the tensors with the indices *i* and *j* has *4 x 4=16* components. The intellectual appeal of this theory, which has continued to fascinate physicists to this day, lies in the link between these abstract geometric objects with the physical concepts of energy and momentum. Einstein's greatest achievement in this context was to stick at it, with intuition and unshakable tenacity. He described his struggle as *"... search in the dark with its tense desire, lasting for years, full of foreboding, with its exhausting change from aspiration to frustration and its final breakthrough to lucidity".[31]*

---

[i] Energy flux density is the energy that flows through a unit surface per unit of time. Momentum is mass times speed. Here too a spatial density (momentum per cubic meter) can be given. Energy flux density and momentum density share the same physical unit. In addition, there is also momentum flux density.

[ii] The so-called Einstein tensor G_{ij} that arises when yet another contraction is added to the Ricci-Tensor.

Einstein's Lost Key

METRICS: HOW FAR IS IT FROM A TO B?

Although Einstein was confident that his equations were correct, a series of simplifications was needed in order to make a specific prediction for observations. They are based on a relatively intuitive concept known as metric.

In 'flat' Euclidean space, the distance between two points is measured according to Pythagoras' theorem: $s^2 = x^2 + y^2$. If the distances involved are infinitesimally small, this is indicated by a prefixed d (for difference): $dx^2 + dy^2 = ds^2$. If the space is curved, however, like the surface of a sphere, then the relevant coordinate (φ, θ) must first be corrected by a factor because the meridian along φ in Scandinavia is shorter than at the Equator!

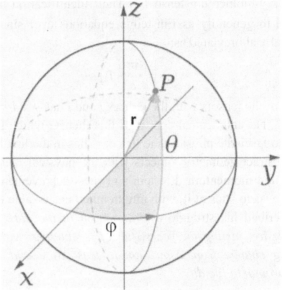

Example of how to calculate distances on Earth, carried out using a 'metric'. Because the geometry is distorted, the spherical polar coordinates (r, φ, θ) are used: $ds^2 = dr^2 + r^2 d\theta^2 + r^2 \sin\theta \, d\varphi^2$. The respective factors at the coordinate differentials $dr^2, d\theta^2, d\varphi^2$ are called metric.

We encounter the same situation when we realize that according to general relativity, the factor by which time passes more

Chapter 6: Seduced by mathematical beauty

slowly in a gravitational field is $\sqrt{1 - \frac{2GM}{rc^2}}$. In space-time a 'rectangle' therefore has sides of different lengths (see fig.).

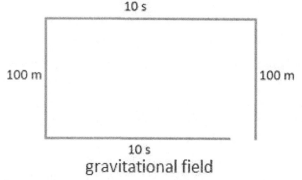

gravitational field

Time passes more slowly in a gravitational field, as measured on numerous occasions. If one considers physical time and space as dimensions in an abstract four-dimensional space, then no conventional rectangle is possible in the 'space-time plane'.

A SOLUTION THAT ACTUALLY SURPRISES

The intriguing property of metric is that its differentiation leads to the (mathematical notion of) connection, and further differentiation[i] to Riemann's curvature tensor. In 1917, mathematician Karl Schwarzschild showed in this way that the gravitational field of the Sun can be described by a fairly simple metric in a way that satisfies Einstein's equations. The calculations are too lengthy to set out here,[32] but the result is worth looking at. As there is spherical symmetry in the solar system, Schwarzschild's metric takes a relatively simple form that involves the radial component r and time t only.

---

[i] To be precise, one would speak of an "exterior" derivative and "covariant" differentiation, i.e. taking curvature into account.

Einstein's Lost Key

$$ds^2 = \left(1 - \frac{2GM}{rc^2}\right)d(ct)^2 - \frac{dr^1}{\left(1 - \frac{2GM}{rc^2}\right)} - r^2 d\theta^2 - r^2 \sin\theta\, d\varphi^2$$

Here we can see that the temporal component of the metric increases by a factor[i] of $1 - \frac{2GM}{rc^2}$, while the units of length are shortened by the same factor. If it is used to calculate the deflection of starlight by the Sun, as Einstein did in 1911, then using the Schwarzschild metric one obtains the value of:

$$\Delta\varphi = \frac{4GM}{rc^2},$$

corresponding to 1.75 arc seconds, twice as much as in Einstein's first attempt. The similarity of the Schwarzschild metric to the 1911 formula is obvious. There, the speed of light had been directly changed, whereas here its variability had been incorporated in time and length scales. The double value (with respect to the earlier 0.85 arc seconds) was not a result of factor 2 in the numerator of the Schwarzschild Coefficient (this is because c^2 is considered rather than c), but rather because in 1911 Einstein had considered only time scales, not length scales. However, the mistake could have been corrected in the variable speed of light approach as well.

At first sight, the Schwarzschild metric will probably not strike you as a model of simplicity, but compared to the 256 components of the Riemann tensor, that's exactly what it is. This astonishing simplification raises the question of whether or not the road taken by Einstein and Grossmann is really the most efficient one in mathematical terms. In particular, formalism as a whole is based on the paradigm that space and time can be treated as equivalent dimensions, a view that shows little re-

---

[i] Note that the term shows up here in square form. The first approximation would be $\left(1 - \frac{GM}{rc^2}\right)$.

Chapter 6: Seduced by mathematical beauty

spect for the speed of light as a natural phenomenon. It has been firmly established since about 1908, and it probably had more fatal consequences than any of the other wrong tracks taken by modern physics.

THE GUIDING PRINCIPLE IS PHYSICS

> *I am now working exclusively on the problem of gravity. One thing is for sure, never in my life have I toiled in a similar way. Smoking a like a chimney, working like a dog, eating without thought or selectivity, seldom going for a walk and sleeping irregularly. - Albert Einstein*[33]

Finally, we must call attention to the structural similarity between Einstein's equations and other formulae in physics. Wherever there is a point-like source in physics, a specific combination of derivatives occurs, called the Laplace operator, which is the analogue of the Einstein tensor. The Laplace operator describes, for example, the electrical potential close to a point charge or the temperature distribution near a source of heat. We shall take a closer look at this similarity in Chapter 8.

One of Einstein's central ideas that led him to his equations was that a tensor is not just a bunch of numbers, but a physical quantity. This meant that he only searched for tensors that could not simply disappear when rotated (generally: under transformations), a quality called as "general covariance". While performing the calculations, Einstein often made mistakes, and without Grossmann's help it would have been hard not to get lost in the jungle of mathematical possibilities.

One specific wrong trail that Einstein and Grossmann had followed turned out to be a particularly unfortunate happenstance. In 1913, they jointly published a preliminary theory they called "draft" ("Entwurf"). Besides mathematical inconsistencies, the calculated light deflection by the Sun was 0.83 degrees – too small. As we know, Einstein had obtained the same incorrect value with his variable speed of light in 1911. Thus even treatments of the theory of relativity that seem historically sound usually lump these two approaches – that have nothing to do with each other – together. The difference between the correct value of 1.7 arc seconds and the incorrect value of 0.83 arc seconds (half as much) is of course important, particularly for experimental verification. But here it conceals the different mathematical strategy of the geometric approach and variable speed of light. The latter description is more apt for revealing fundamental relations to the universe. Even in the variable speed of light context, the deflection was correct – but this was to remain undiscovered for almost fifty years.

THE ROAD TO FAME

The scientific community eventually began to take an interest in the works of Einstein and Grossmann on general relativity. The famous mathematician David Hilbert became aware of them through a series of lectures that Einstein gave in Göttingen in the summer of 1915. Hilbert's thinking wasn't really physical, but deriving transformation properties of tensors was a piece of cake for him. Einstein suddenly became afraid that someone might snatch away the discovery he hoped to make. Hilbert had possibly submitted "Einstein's equations" for publication a little earlier than Einstein, and to this day it is not clear who got there first.[34] There were also resentments in the private correspondence between Einstein and Hilbert. Eventually Hilbert had the generosity to concede that general relativity, as a physical theory, was Albert Einstein's alone. Even today historians assess events in November 1915 differently, and how exactly the final shape of the geometric version of general relativity came about is the subject of continuing interest. This is also one reason why Einstein's 1911 idea of variable speed of light fell into oblivion.

Chapter 6: Seduced by mathematical beauty

Einstein himself was firmly convinced of the correctness of his calculations at this time. In his memoirs he wrote that he was too excited to be able to sleep when he obtained the correct value for Mercury's perihelion shift – an observational anomaly known to astronomers since 1859 that Newton's law could not explain.[i] Yet despite the perihelion shift, the theory was not generally accepted. However, the events in November 1915 prompted scientists to check the theoretical predictions by observation. The world was eager to see whether Einstein's calculations would prove to be correct or not.

German astronomer Erwin Freundlich had tried to confirm the theory at an early stage. In August 1914, he had undertaken an expedition to the Crimea. The expected eclipse in this region should have helped Einstein's theory to make its breakthrough, but the outbreak of the First World War interfered. The researchers, with their suspicious-looking instruments, were arrested as spies. If Freundlich had measured the deflection of starlight by the sun at that time however, people would probably have considered Einstein's theory a failure, because in 1914 he still believed in his incorrect prediction of a deflection of 0.83 arc-seconds.

Not until the First World War had ended were conditions in Europe once again such that science could flourish. Sir Arthur Eddington, a young physicist, Quaker and conscientious objector, set about checking Einstein's prediction – even though it came from a country that had been at war with his own for four years.

---

[i] Here too Einstein's opponents accused him of plagiarism, as Berlin physicist Paul Gerber had published a formula on the perihelion shift in 1898 (Z. Math. Phys. 43, pp. 93–104). Einstein wasn't familiar with the approach – which lacked a coherent justification, incidentally. It has to be conceded, however, that Gerber's formula has a certain heuristic value.

SHOWDOWN IN LONDON

Eddington organized expeditions to the remote areas of Sobral (Brazil) and Principe (Western Africa), which had to grapple with unlucky circumstances. The weather was persistently overcast, making photography almost impossible – and when some pictures were eventually taken, it turned out that the heat had damaged the photographic plates. This made it technically difficult to evaluate the results, and initially they were not as unambiguous as Eddington had wished, almost contradictory. He decided to leave out some photographic plates taken by a supposedly faulty instrument. Then, the analysis favored the interpretation of a light deflection of 1.7 arc seconds.[35]

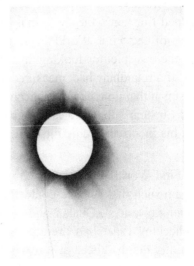

Photograph of the solar eclipse in Principe on May 29th, 1919 (photographic negative).

Finally Eddington presented his results in a memorable meeting of the Royal Society on November 6th, 1919. England was the leading physics nation at the time, and a long list of celebrities had accepted Eddington's invitation – one of them the 84-year-old Lord Kelvin, who had significantly influenced the de-

Chapter 6: Seduced by mathematical beauty

velopment of physics as a whole towards the end of the 19th Century.

Although the data alone would still have justified residual doubts, and the exclusion of some photographic plates should have been discussed, Eddington's message was clear: he stated that the measurements vindicated Einstein's new theory of gravitation. One has to imagine the atmosphere of this lecture and its impact on the physics world, in a time-honored hall, overlooked by a bust of Newton. Many of those of the old school did not agree, and some of them, like Lord Kelvin, actually walked out. However, the overall atmosphere was enthusiastic.

Practically speaking, this moment marked the widespread acceptance of general relativity. From a historical and sociological perspective this is remarkable, because apart from Eddington nobody present could have checked the theory properly. The anecdote about Eddington doubting whether more than two people understood general relativity supposedly happened that evening. Thus, evidently, the establishment of a scientific discovery has a considerable sociological component.

On the one hand, the venue lent the measurements a degree of credibility that in subsequent years (by a variety of observations, including modern high-precision tests) became justified. Much more questionable however was the "establishment" of the theory as a unique description of the observations. Although the theory has withstood all experimental tests so far, the meeting in 1919 had fatal consequences: that the geometric version of general relativity may not be the only one to explain the measurements was overlooked. Instead of resorting to curvature of space, light deflection could also find its explanation in variable speed of light.

ARRIVAL IN THE NEW WORLD

The next day, the *New York Times* printed the headline "Einstein theory triumphs", adding more prosaically "Stars not

Einstein's Lost Key

where they seemed ... to be, but nobody need worry". The scientific sensation was complete. Einstein became a celebrity that day, providing the geometrical formulation of general relativity with a cloud of incomprehensibility – which made Einstein even more famous. Ever since that day, it has been unthinkable that the results could be presented more simply and more intelligibly.

LIGHTS ALL ASKEW IN THE HEAVENS

Men of Science More or Less Agog Over Results of Eclipse Observations.

EINSTEIN THEORY TRIUMPHS

Stars Not Where They Seemed or Were Calculated to be, but Nobody Need Worry.

A BOOK FOR 12 WISE MEN

No More in All the World Could Comprehend It, Said Einstein When His Daring Publishers Accepted It.

New York Times issue dated 7 November, 1919. "Stars not where they seemed ... to be, but nobody need worry. – A book for 12 wise men"

Chapter 7
Gravity From the Universe

Einstein could not see that Mach was correct

The geometric version of general relativity has been unshakably established ever since the events of 1919. Given that the true size of the universe would not be revealed until a decade later, this can only be seen as a tragic sequence of events – because the cosmological observations of the 1930s would have given a clear hint that Einstein was on the right track with his 1911 attempt. The key to this problem is Newton's gravitational constant G, which Einstein included in his general theory of relativity. Perhaps he ought not to have done so. We shall now take a closer look at the riddle surrounding G.

Isaac Newton once poetically phrased the nature of the gravitational force: "The reasons lies in the property of all masses to attract each other." More prosaically, the power of gravitation is simply proportional to the product of two masses – for example the mass of the Sun (M) and the mass of the Earth (m). Eventually Newton realized that this mysterious attractive force decreased in proportion to the square of the distance between them (r). This means that a planet twice as far away from the Sun as the Earth would be affected by only a quarter of its force.

If we want to calculate the amount of force however, one element is still missing: Newton's so-called gravitational constant G, which completes the law $F = \frac{GMm}{r^2}$. From celestial observations, only the numerical value of GM (the so-called Kepler constant), could be determined, but neither the value of G nor the masses of celestial bodies. In fact Newton believed that because the value of G was so tiny (mind that celestial bodies are enormously heavy) this would never be possible.

It was not until a century later in 1798 that Henry Cavendish was able to measure the value of the gravitational constant in a legendary experiment. The idea behind the experimental apparatus, which is still in use today, goes back to John Michell (1724–1793), a truly visionary natural philosopher. He made another contribution to the theory of gravitation that we shall consider later.

THE NUMBER PUZZLE

Today's best measurements vary around the value $6.673 \cdot 10^{-11} \frac{m^3}{s^2\,kg}$, about which there has been a great deal of discussion in recent years.[36] The number is a mystery. Why *6.673?* Why not *5.334?* Contrary to the claims of adherents of the 'Anthropic Principle' (one of the esoteric sicknesses of present-day physics), this specific value was not indispensable for the origin of life. If *G* was twenty percent smaller, nobody would be bothered – and our vertebral discs would be grateful to us. If physics is understood in a philosophical manner as questions put to nature, then something unsatisfactory remains in such a numerical value. In a 1953 essay[37], cosmologist Dennis Sciama summed it up that:

> *"Newton's gravitational theory contains two arbitrary elements – the choice of the coordinate system and the value of the gravitational constant."*

We covered the first point in detail in Chapters 4 and 6. Einstein elegantly solved the problem of accelerated reference

Chapter 7: Gravity from the universe

frames[i] using tensors that maintain a physical significance in any coordinate system. What remained unsolved, however, was the issue of the gravitational constant – it isn't even clear whether Einstein had seen it as a problem by 1915. We should not forget that the completion of general relativity and his later reflections on fundamental constants were separated by thirty years. The Einstein of 1915 may not have pondered deep questions of natural philosophy, as his later self did. Or perhaps he was so intoxicated by success and enthusiastic recognition that he disregarded the seemingly small shortcoming of which he became aware in later years.

The context of tensor analysis that we touched on in the last chapter is also conceptually so remote from reflection on constants of nature that it would have been difficult for him to deal with both problems at once. Wherever his thoughts may have been, neither in 1911 nor in 1915 was Einstein aware that the gravitational constant was an arbitrary element whose elimination was desirable. Had Edwin Hubble made his discovery about the universe twenty years earlier, Einstein could not have failed to notice it. From today's point of view, it would be almost trivial to recognize the connection between G and the size of the universe. Why?

NATURE SPEAKS

The simplest way to approach the problem is to look at the units of the gravitational constant ($\frac{m^3}{s^2\, kg}$), which are already set in stone by Newton's law of gravity – because the left-hand side of the equation $F = \frac{GMm}{r^2}$ must necessarily result in a unit of force, the Newton (N).[ii] The occurrence of s^2 in the denominator

---

[i] We shall come back to another approach to this first problem, which may be more closely linked to Ernst Mach's principle.
[ii] $N = kg\, m/s^2$ as in $F=ma$, always holds.

and m³ in the numerator suggests that the speed of light squared is somehow contained in G. The residual unit m/kg, almost cries for a cosmological interpretation in which the size (in meters) and mass (in kilograms) of the universe occur.

Around 1911, however, these quantities could not be measured, unless one confused our comparably tiny galaxy with the cosmos as a whole. Nor did anybody have any idea that this cosmos would be discovered at all. Since such a quantitative test of the connection was not even remotely in sight, Einstein refrained from speculating about it – in 1911 at least. It is fairly certain, as we shall see below, that Einstein later touched on a similar formula, although in a context that probably blocked his view of the most interesting aspect.

Let us be clear: no cosmology existed in 1911. But the mass and size of the universe are obviously contained in the gravitational constant! The visible mass of all the galaxies in the universe adds up to approximately 10^{53} kilograms, while its size can be fairly estimated to 10^{26} meters. Dividing one by the other yields 10^{27} kilograms per meter, coinciding with the value derived from the gravitational constant, c^2/G. Of course the huge uncertainties in astrophysical observations apply to these figures, but all measurements, including those made with present-day precision (which still require interpretation) have come up with the same answer. In a word, $G \approx c^2 \frac{R_u}{M_u}$ is true. That is sensational.

It is hard to believe that such an obvious relation of units and numerical values can be coincidental. Yet the formula has not triggered a revolution as the comparable case with $1/c^2 = \varepsilon_0\mu_0$ in electrodynamics. How it could have happened that generations of physicists failed to notice the obvious? This will certainly someday become a subject for the sociology of science. It is as clear as day that Ernst Mach's idea – that masses determine the gravity – shows up here.

Chapter 7: Gravity from the universe

One cause of this blindness is the superficial use of the term 'energy' in a cosmological context. The conservation of energy is generally said to be a law of nature. Few are aware instead that energy was invented precisely for this: a quantity that does not change with time during all dynamical processes. Yet the universe may change so slowly (in comparison to its enormous age) that it is easy to overlook. Therefore, there is no reason at all to assume that energy is conserved over cosmological periods of time. However, the enigmatic formula $G \approx c^2 \frac{R_u}{M_u}$ can also be interpreted (in this superficial view) as the equivalence of kinetic and potential energy in the universe. The total energy would then always be equal to zero – but there is definitely no reason for that. Marveling at that coincidence instead of questioning the origin of G is tantamount to the behavior of a dog chasing its own tail.

Nowadays, the coincidence as such is indeed not unknown, but is described using a bizarre theory of "cosmic inflation". This implies a God-given gravitational constant decorated with hypotheses – however arbitrary they might be – about the universe's first micro-fractions of a second after the Big Bang. We are told that the universe has expanded at multiple times the speed of light while creating a plethora of 'bubble universes'. Go figure. This so-called theory of cosmic inflation has been exposed as absurd by some sane physicists,[38] but it still gains more and more traction within the "community". But this is not the place[39] to analyze the lunacies to which groupthink may lead.

SCHRÖDINGER'S HOUR OF GLORY

There is a real gem of physical reasoning in a completely unknown article[40] on cosmology published in 1925 by Erwin Schrödinger, who was later awarded the Nobel Prize. Today he is best known for his essential contribution to quantum mechanics: the wave equation that bears his name, which he found incidentally, also in 1925 (during a skiing holiday in Switzerland with a lover who remained unidentified). Schrödinger's thoughts on cosmology are perhaps no less important, even though they are entirely forgotten. He, in fact, was the first to suspect the coincidence $G \approx c^2 \frac{R_u}{M_u}$!

Erwin Schrödinger (1887–1961)

Whereas the relation $G \approx c^2 \frac{R_u}{M_u}$ as such is only numerical, Schrödinger went one step further and realized that the concept of the gravitational potential φ was concealed in the formula. Potential is simply energy per mass, for which Newton had

Chapter 7: Gravity from the universe

derived an expression in his theory of gravitation: $\varphi = -\frac{GM}{r}$, when a mass is at a distance r from the Sun (with mass M).[i]

> **8. Die Erfüllbarkeit der Relativitätsforderung in der klassischen Mechanik; von E. Schrödinger.**
>
> Gegen die klassische Punktmechanik mit Zentralkräften, deren Grundlagen in klarster Form von L. Boltzmann[1]) herausgearbeitet wurden, ist bekanntlich schon von E. Mach[2]) der Einwand erhoben worden, daß sie der vom erkenntnistheoretischen Standpunkt sich aufdrängenden Relativitätsforderung nicht genüge: ihre Gesetze gelten nicht für *beliebig*

First page of Schrödinger's original article of 1925: "How to fulfill the postulate of relativity in classical mechanics." Mach's principle is mentioned.

Let us point out for a moment the subtle difference from gravitational force $F = \frac{GMm}{r^2}$, where the distance is squared in the denominator. This means that the gravitational force for distant celestial bodies strongly decreases, and the gravitational force the Sun exerts on the Earth is thus hardly noticeable (apart from the tides, to which it contributes). The gravitation potential is quite a different matter: the value of the solar potential in which we find ourselves exceeds the effect of the Earth by a factor of ten – which is easy to see if one considers the two quotients M/r (mass divided by distance).

Schrödinger noticed that too. It looked plausible to him that the influence of the even more distant masses in the Milky Way had to be larger, even though it was impossible to perceive a force. Schrödinger tried to estimate this potential and noticed, of course, that it had the same unit as the square of the speed of

---

[i] To obtain energy, the term must then be multiplied by the mass m.

light, c^2. With amazing intuition he suspected that all the potentials in the universe might just add up to c^2. In Schrödinger's own words (p. 331):

> *"This remarkable relationship states that the (negative) potential of all masses at the point of observation, calculated with the gravitational constant valid at that observation point, must be equal to half the square of the speed of light."*[i]

In spite of the rudimentary astronomical data back then, he concluded that this indicated a far bigger universe than it was known at the time (p. 332):

> *"Thus only a vanishingly small fraction of the inertial effect observed on Earth and in the solar system can originate from their interaction with the masses of the Milky Way."*

In a way Schrödinger had thus anticipated the discovery of the size of the cosmos in the 1930s. He further insisted that Mach's principle had to be incorporated into the theory of relativity. In this respect, Schrödinger's intuition went beyond Einstein's. This makes it all the more bizarre that Schrödinger's work on cosmology is completely unknown even among physicists.

EINSTEIN'S UNKNOWN CALCULATION

There remains, however, the highly interesting question of whether Einstein knew of the coincidence $\frac{c^2}{G} \approx \frac{M_u}{R_u}$. Even if he had not fully formulated this idea, did he at least ever think about it? Apparently he did! Neither in Einstein's publications nor in his letters is this documented, but a remarkable passage is

---

[i] The factor ½ is not significant for the estimate.

Chapter 7: Gravity from the universe

found in the memoirs of Alexander Moszkowski[i] (a really good historical read, incidentally) about their discussions. Moszkowski recounts:

> "Einstein succeeded in determining the approximate size of this non-infinite universe. He deduced that from the existence of a measurable gravitational constant.... [ ]. He further assumes that the distribution of matter corresponds to the average density of the Milky Way. As a result of the calculations, Einstein obtains the following quantity: He finds that the whole universe has a diameter of about 100 million light years."

This is a stunning statement. Evidently, Einstein used a value for the density (mass per volume) of the Milky Way estimated by the astronomers of the time[ii] in order to calculate the size of the cosmos. However, it is impossible to derive the size (R_u) from the density ρ without additional information. Since ρ is proportional to M/R_u^3, thus Einstein must also have assumed the coincidence $\frac{c^2}{G} \approx \frac{M_u}{R_u}$. If we combine the two formulae, then Einstein's estimate mentioned by Moszkowski could only have been $R_u \approx \sqrt{\frac{c^2}{G\rho}}$. Using the density of the Milky Way established at the time (which was almost a million times greater than that of the universe), the result matches the value quoted by Moszkowski. Thus Einstein must have been aware of the relation $\frac{c^2}{G} \approx \frac{M_u}{R_u}$ even if he never mentioned it!

---

[i] Brother of the composer Moritz Moszkowski.
[ii] One historical source for this is the aforementioned 1925 article by Erwin Schrödinger. He writes (p.323): ".... uniformly filled with stars of the mass of the sun Such that 30 such stars fill a sphere with a radius of 5 parsecs." (1 parsec is equal to 3.26 light years or 3.08 x 10^{16} meters). Using Schrödinger's estimate, the result for R_u is about 60 million light-years instead of 100.

It is not easy to date Einstein's insight that Moszkowski had written down around 1919. Moszkowski refers to a report of the meeting of the Prussian Academy of Science[i] of February 8th, 1917, but he adds that Einstein had not explained the idea yet. This relatively well-known 1917 article "Cosmological considerations on general relativity" contains only indirect references to the above relation.

The context of the article – further discussed below – suggests, however, that Einstein's interpretation of $G \approx c^2 \frac{R_u}{M_u}$ here was very different from Schrödinger's. Apparently Einstein did not consider the gravitational potential, and made no clear reference to Mach's principle even though in Schrödinger's account this leaps to the eye. The formula that links the gravitational constant to the gravitational potential of distant masses is nothing other than a quantitative statement that gravity originates from the distant masses in the universe. At this point, Einstein failed to draw the big picture.

EINSTEIN APPRECIATED MACH...

This is all the more regrettable, as Einstein's regard for Mach is well documented. Unfortunately they met only once, in Vienna in 1911. Mach was already hard of hearing, and they engaged in a debate about atoms rather than about the origin of gravitation (which perhaps would have been more fruitful).

In 1913, Einstein wrote a very kind and respectful letter[41] to Mach, stressing that that Mach's "ingenious investigations" would be "brilliantly confirmed" if his theory of general relativity was shown to be correct.

---

[i] http://echo.mpiwg-berlin.mpg.de/ECHOdocuView?url=/permanent/echo/einstein/sitzungsberichte/S250UZ0K/index.meta&start=1&pn=10

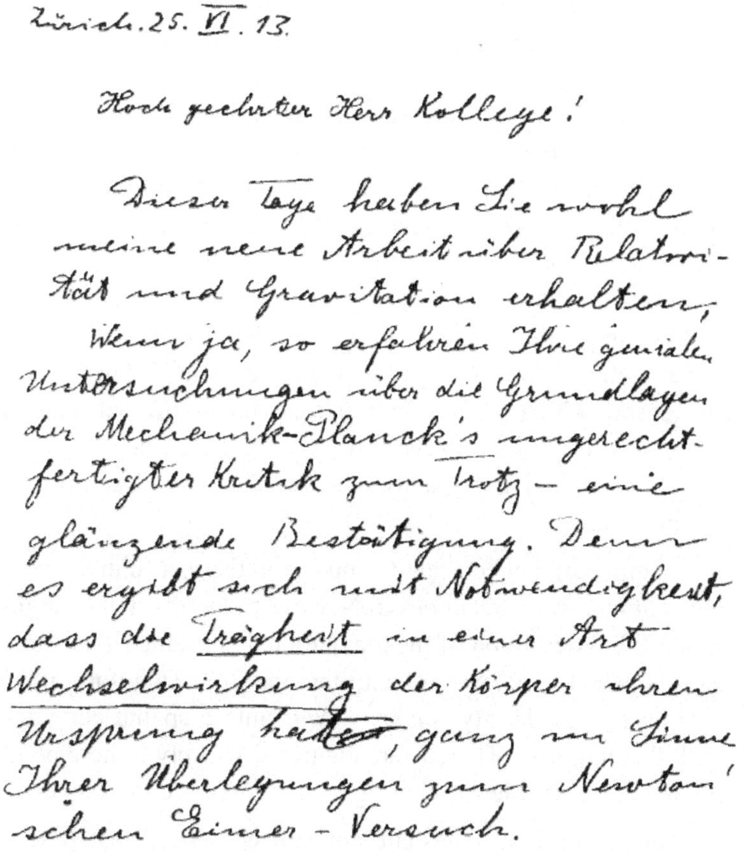

Chapter 7: Gravity from the universe

[Handwritten letter in German:]

> Zürich. 25. VI. 13.
>
> Hochgeehrter Herr Kollege!
>
> Dieser Tage haben Sie wohl meine neue Arbeit über Relativität und Gravitation erhalten. Wenn ja, so erfahren Ihre geniale Untersuchungen über die Grundlagen der Mechanik-Planck's ungerechtfertigter Kritik zum Trotz — eine glänzende Bestätigung. Denn es ergibt sich mit Notwendigkeit, dass die Trägheit in einer Art Wechselwirkung der Körper ihren Ursprung hat, ganz im Sinne Ihrer Überlegungen zum Newton'schen Eimer-Versuch.

Excerpts from a letter from Einstein to Mach dated June 25th, 1913. He writes about the paper he had sent to Mach, and claims that it confirms Mach's thoughts 'brilliantly', referring to the rotating bucket thought experiment.

Mach felt however rather misunderstood than honored. Shortly after, with apparent reference to that letter, he wrote:

> "I learn from the publications I am receiving, and in particular from my correspondence, that I am going to be intended to take the role of a trailblazer of the theory of relativity. I can now get an idea of how the thoughts set out in my

Einstein's Lost Key

> *'mechanics' will undergo construal and reinterpretation from this site."*

This was certainly a sarcastic exaggeration by Mach, who was not an easy person – but it is true that Einstein's understanding of Mach's principle was more limited than Mach had intended. In an article[42] in the *Annalen der Physik* Einstein claimed that the "G-field" had to be determined by the masses, and tried to contrast himself with "expert colleagues" who "do not see the need to implement Mach's Principle". However, by "G-field" Einstein did not mean the gravitational constant G, but the metric tensor, which of course is a function of the surrounding masses as in the case of the force law. But this did not implement Mach's idea that gravity was caused by the distant masses out in the universe.

The gravitational constant G appears in the metric in a merely conventional way, as it does in Newton's law. Einstein's slightly modified definition of his gravitational constant κ ("kappa"), as it is called, is $\kappa = 8\pi G/c^4$ with the unit $1/N$. Therefore, it can turn an energy density $Nm/m^3 = N/m^2$ into a spatial curvature[i] with the unit $1/m^2$. There is an additional subtlety in here pointing towards variable speed of light, as we will see in Chapter 8.

Although in 1912, Einstein was still convinced that Mach's principle was essential, he later conceded with regard to the geometric formulation of 1915:

> *"Of course I had come to know Mach's point of view, according to which it seemed possible that inertia was not a resistance to acceleration as such, but the consequence of an acceleration with respect to the masses of all other bodies present*

---

[i] Carl Friedrich Gauss first defined the curvature of a surface with two opposing, perpendicular circles whose arcs cling to the surface. The curvature is then the reciprocal of the product of the two radii.

Chapter 7: Gravity from the universe

> *in the world. There was something fascinating about this idea but it offered nothing useful as a basis for a new theory."*

... BUT LEFT HIM OUTSIDE

Contrary to Einstein's suggestions and despite the obfuscation[i] laying on the topic to this day, Mach's principle is – unfortunately – not part of general relativity, at least not completely.[ii] A brief thought experiment makes this clear:[43] imagine a universe in which only two masses orbit each other, driven by mutual attraction. In the conventional theory this would make no difference, because the gravitational constant G is always the same. But if we had it calculated by $G \approx c^2 \frac{R_u}{M_u}$ or by Schrödinger's formula, the attraction between these two "planets" would be much stronger – comparable to nuclear forces inside atoms! This is another mysterious coincidence, which we shall discuss in Chapter 9.

Anyway, Einstein would probably have incorporated this relation (which he had touched upon in a different context around 1917) directly into his 1911 theory, had he been aware of the true size of the universe back then. The history of cosmology would have taken a different course.

Obviously, Einstein was not completely satisfied with the cosmological consequences of his 1915 theory. In fact, around 1917 (a year after Mach's death) he discovered that his field equations grown out of the geometric formulation did not permit a static universe. In order to rescue his preferred model, Einstein added a "cosmological constant" Λ (pronounced: "lambda") to his equations, which made the universe static

---

[i] The so-called Lense-Thirring effect, for example, is incorrectly identified with Mach's principle.
[ii] Very clearly also stated by Schrödinger in 1925.

again but suffered from a serious drawback. A cosmos with Λ would be unstable: even a tiny disturbance would make the universe explode or collapse.

This failed model apparently prompted the calculation Moszkowski had referred to. In order to give a numerical estimate for his Λ, Einstein used the data of the Milky Way. In this rather awkward construction, the value $r = \frac{GM}{c^2}$ was vaguely called the "curvature radius" of the universe, a term that is nowadays known as the "Einstein radius". According to Schrödinger's much simpler interpretation, it is in fact the radius of the universe itself![i]

> *Although the researcher is happy to reach for what is most immediately attainable, an occasional glance into the depths of the unknown certainly does him no harm. – Ernst Mach*

MACH DIES A LONELY MAN

At the time, debates in the literature focused on the question of whether the universe was static or expanding. The arguments were predominantly formal, concerning so-called boundary conditions of the universe at an infinite distance. The far more important problem of how to realize Mach's principle faded away. Schrödinger was much annoyed by the ivory tower math of his colleagues dabbling in cosmology, who deliberately ignored Mach. In 1925, he wrote:

> *"Because of the conceptual difficulties that still encumber these cosmological theories, and not least because of the mathematical difficulties in understanding them, a solution of a paramount*

---

[i] This essentially means the distance from which light has been traveling since the Big Bang to reach us. This definition retains its validity in the modified cosmology described in Chapter 10.

Chapter 7: Gravity from the universe

> *epistemological question that is obvious to all erudite people was shifted to a field where few can follow, and where it is not easy to distinguish clearly between truth and poetry."*

Einstein's 1917 model with Λ was the beginning of a long series of cosmological models, which to this day try – and fail – to adapt to reality, despite their continuous fine-tuning. Again, this strongly indicates that the geometric version of general relativity, albeit successful in the solar system, is a faulty design when viewed from a cosmological perspective. In 1929, when Edwin Hubble – Vesto Slipher and Georges Lemaitre should not go unmentioned here – finally discovered the red shift, Einstein's pet model of a static cosmos faced a major problem.

Based on Einstein's equations, Dutch astronomer de Sitter and Russian mathematician Friedmann had previously developed a model of the cosmos that conformed to the new interpretation as expansion. This so-called Einstein-DeSitter model, which could still claim a certain simplicity among the available choices, was favored for a long time. Eventually, it had to be abandoned because it conflicted with the observations that are now referred to as "accelerated expansion". We shall discuss these modern observations in Chapter 11.

FATAL ADMIRATION

In retrospect, it is hard to understand why Einstein allowed himself to be drawn by his contemporaries into debates on a series of ultimately irrelevant cosmological models, rather than recalling his own 1911 proposal. Apparently, too many years had passed, but the major reason was that around 1915, Einstein had convinced himself – as thoroughly as wrongly – of the inadequacy of his earlier attempts. He seemed to have forgotten the idea of variable speed of light, which could have given a much simpler and more accurate picture of the universe. Contact with the other researchers hot on his heels certainly did the

Einstein's Lost Key

loner Einstein no good in his quest for the laws of nature. For, in one respect, his biographer Abraham Pais, in his analysis of 1911, was right when he wrote: "Science was his life, his devotion, his retreat and the source of his detachment. He was completely alone in Prague." Einstein's celebrity status from 1915 onward helped to distract him from his best idea.

However brilliant those elite mathematicians and astronomers may have been, they were in first place copycats of Einstein's theory of relativity. As soon as they had learned to deal with its mathematical difficulties, they uncritically accepted it as the truth. Believing in Einstein's authority, they encouraged him to hold on to the geometric version, the scientific survival of which would have been far less important to Einstein than to his followers. Many of them – Friedmann, Lemaitre, de Sitter and others – would today be just as forgotten as Max Abraham if Einstein had returned to his theory of variable speed of light.

Instead of turning into the long agony of model fiddling that continues to this day, Einstein's theory of 1911 was within an ace of becoming a triumph of cosmology. Another opportunity showed up fifty years later, as we shall see in the next chapter.

Chapter 8

Half a Century Too Late

Robert Dicke's simple completion of Einstein's idea

It is not unusual for great names to become famous for discoveries that are in fact inferior to their other works. American astrophysicist Robert Dicke (1916-1997) is an eminent example of this category. Dicke is of course well known among physicists, but hardly anybody would put him on a par with Schrödinger, Dirac and Einstein. I would like to suggest that this might be a mistake.

Robert Henry Dicke (1916-1997)

Einstein's Lost Key

Dicke was enormously versatile. He was a dedicated experimentalist, and his contribution to the detection of the signal known as cosmic microwave background is legendary. While his group was working on a suitable antenna, it turned out that other scientists had already made the measurement ("Boys, we've been scooped"). They had no idea of the value of their discovery, which Dicke was the first to explain to them – but they were awarded the Nobel Prize for it. Apart from his contribution, his attitude then was enormously decent compared with what is usual in today's competitive environment.

Even though, around 1969, he was the first to think about the baffling relationship $c^2 \approx G \frac{M_u}{R_u}$ (known as 'flatness' today) in a completely new way, Dicke's theoretical works received little attention. His calculations made the relation between the gravitational constant G and the universe seem even more mysterious. It is fair to say, however, that Dicke showed that it could not be just coincidence. Later theorists of 'cosmic inflation' have hijacked Dicke's observation and claimed that inflation explains it. Dicke was only mentioned parenthetically by the guru of inflation, Alan Guth, and went unnamed in the ensuing publications.

Finally, Dicke developed a variant of general relativity, which in the 1960s attracted a certain amount of attention and is now known as 'scalar tensor' or Brans-Dicke theory. Far less well known is the path Dicke took to get there, which is far more interesting than the 1961 theory itself. Dicke held in his hands the key to a cosmological revolution that Einstein had lost.

In 1957, Dicke published an article[44] entitled "Gravitation without a principle of equivalence," which looks fairly abstract at first glance – and this may explain why little notice was taken of it. The content is sensational. It concerns Einstein's theory about variable speed of light dating back to 1911, but in contrast to Einstein, Dicke obtained the correct light deflection. While doing this he managed two great feats: correcting Einstein's

Chapter 8: Robert Dicke's simple completion

mistake, which had held back the spread of the theory, and even more importantly, incorporating Mach's principle that was so revered by Einstein. Let's take one at a time.

REVIEWS OF MODERN PHYSICS   VOLUME 29, NUMBER 3   JULY, 1957

Gravitation without a Principle of Equivalence

R. H. DICKE

Palmer Physical Laboratory, Princeton, New Jersey

THE previous article has considered the observational and experimental facts and has concluded that there is no substantial evidence to support the belief that the coupling constants of the weak interactions are independent of time or place. Consequently, nates, the curvatures of a space are modified. With a proper redefinition of units making them dependent upon coordinates and orientation of an infinitesimal interval a curved space can be converted into a flat one and vice versa. Rosen[1] has shown how to formulate

Title page of Dicke's original publication in 1957.

WHAT EINSTEIN HAD FORGOTTEN ...

Dicke starts his discussion with an insightful crosslink between the special theory of relativity and continuum mechanics,[i] in that the speed of sound (analogous to the speed of light) can also vary. Ultimately, Dicke saw the same parallel with optics, developing – as Einstein did – a formula for the ratio between the speed of light close to the sun (c) and the speed far away from it (c_0). As in optics, Dicke called this a refractive index n (while Einstein had given a formula for $1/n$ which makes, as we shall see, a subtle but important difference). Dicke's formula for the refractive index was therefore

$$n = \frac{c_0}{c} = 1 + \frac{2GM}{rc^2}$$

The major difference from the formula that Einstein had considered half a century before is the factor of two on the right, which ensures that the value obtained for the deflection of a ray of light is not just 0.83 arc seconds but in fact the double value, about 1.7 arc seconds. That's all very well, one might say, and

---

[i] This, in its turn, is related to the Einstein-Cartan theory of 1930 and theoretical findings in continuum mechanics in the 1950s (see e.g., my comments in arxiv.org/abs/gr-qc/0011064)

of course it was easier for Dicke because he was, unlike Einstein, aware of the solar eclipse observations. The aforementioned difference in the formulation of the refractive index resulted in Dicke's writing $n=1+x$ (where x is a small number), whereas Einstein chose the form $1/n=1-x$. Since $0.998 \approx \frac{1}{1.002}$, this appears to make very little difference at first sight.

Dicke's great accomplishment was that he tried to make sense of the observations without having too much respect for the conventional formalism of the theory. He had the courage to come up with an alternative to general relativity. Yet how did he arrive there? Einstein, when dealing with variable speed of light, considered only clocks running at different speed, that is, variable frequencies f. Dicke, in addition, realized that in a gravitational field, not just clocks run slower: wavelengths produced by atoms are shorter, too. This means that in the formula $c = \lambda \cdot f$ the decrease of c in the gravitational field is shared equally by the frequencies f and wavelengths λ. If we denote the corresponding differences with the Greek letter delta (Δ), then:

$$\frac{\Delta c}{c} = \frac{\Delta f}{f} + \frac{\Delta \lambda}{\lambda},$$

which is essentially is the same as the above

$$\frac{f}{f_0} = \frac{\lambda}{\lambda_0} = \left(1 + \frac{GM}{rc^2}\right)$$

The decrease might be written symbolically as $c\downarrow\downarrow$ $\lambda\downarrow$ $f\downarrow$, where the arrows indicate a relative decrease. Dicke solved the major problem of Einstein's 1911 theory almost casually.

But is it really that easy to formulate the intricate concept of curved space with a variable speed of light? Yes, it is. Curvature is nothing other than curved rays of light. They search for the fastest path, not the geometrically shortest. Wherever the quickest path is not a straight line, there is curvature. Every time Google Maps shows us the fastest route we have an analogous situation: the very same optimization problem that a ray of light

Chapter 8: Robert Dicke's simple completion

has to solve. At each location (highway, road, cross-country), there is a different speed, and therefore it saves time not to take the direct geometrical path. Generally speaking, whether we call it the fastest path in flat space or the shortest path in curved space makes no difference. But this insight only sinks slowly into the brains of theoretical physicists, who have long been accustomed to curved space.

... AND WHAT WOULD HAVE EXCITED EINSTEIN

Dicke's use of the positive sign in his formula $n = \left(1 + \frac{2GM}{rc^2}\right)$ led him now to an intriguing observation. He wrote: "Obviously, the second term $\frac{2GM}{rc^2}$ is small compared to 1, and it is due to the presence of the Sun. But what about the first term, 1? Could it originate from the remainder of matter in the universe?"[45]

The simplicity of this bold idea makes it immediately compelling, and since the relation $c^2 \approx G\frac{M_u}{R_u}$ is within observational accuracy, Dicke's assumption was correct! He recalculated and commented:[i] "From the point of view of Mach's principle, this is a highly satisfying result." Not only had Dicke realized Mach's vision (inertia and gravitation are owed to the presence of all masses in the universe) in a concrete formula, he also incorporated it with Einstein's general theory of relativity.

I fail to understand why this brilliant result has not received due attention. I can only assume that because Dicke, at that time, was still a relative unknown a fundamental revision of general relativity was not expected of him. The accomplishments that brought him to fame all came later, and the fact that his 1957 work did not ignite a revolution in physics was, unfor-

---

[i] In 1925 Schrödinger had referred to Mach in very similar terms. Dicke, of course, knew nothing of Schrödinger.

tunately, his own fault. As if the tragedy of 1911 was destined to repeat itself, Dicke committed at least one error that let his results seemingly disagree with the measurements. Finally, there was one simple reason why Dicke did not mention that he had improved Einstein's 1911 theory: he did not realize it! Although Dicke did not mention Einstein, it seems almost impossible to me that he had considered a formula so similar to Einstein's without knowing it. Dicke even referred to Ernst Mach, after all.

PEARLS OF THOUGHT

However, in e-mail correspondence I had with Dicke's colleague Carl Brans, Brans expressly confirmed that Dicke had no idea that he had followed the same approach as Einstein.[i] I am sure that, had Dicke mentioned Einstein in his 1957 article, discussion would have arisen.

In this book we try to have a look at the historical circumstances behind the facts, and thus we still have to render justice to one researcher who, similarly to Einstein, Schrödinger and Dicke, had established the relation between the speed of light, the gravitational constant and the universe: the British-Egyptian cosmologist Dennis Sciama (1926-1999). In 1953, he wrote an interesting article "On the origin of inertia."[46] In the spirit of Ernst Mach, he suggested a formula[ii] for G that may be regarded as a further development of Einstein's formula:

---

[i] Despite their names being linked to a well-known concept, Carl Brans and Robert Dicke seemed to be an imperfect match. Brans wanted to write a thesis on theoretical physics, preferably with Charles Misner, who sent him on to Dicke. This is akin to a composer wanting to study with Salieri who is then disappointed that he got involved with Mozart.

[ii] Schrödinger wrote this as an integral, not as a sum with endless small terms of the sum, which however alters nothing in principle. Equally irrelevant is the fact that this, in accordance with the definition, appears in the publication with a negative sign.

Chapter 8: Robert Dicke's simple completion

$$G\frac{m_1}{r_1} + G\frac{m_2}{r_2} + G\frac{m_3}{r_3} + \cdots = \sum_i G\frac{m_i}{r_i} = c^2$$

Mathematicians, who are always lazy about keeping a record, abbreviate a sum such as the many terms on the left with a Greek Σ (pronounced: "sum over all i..."). Sciama's sum contained all gravitational potentials and also the suggestion that this sum matched the value of c^2.

Dennis Sciama (1926-1999)

However, Sciama was concerned with G rather than with a variability of c. Furthermore, he considered his formula an approximation, and in order to agree with Dicke another factor 4 is required.[i] It is however fascinating that another thinker independently arrived at the same conclusions. Sciama, in 1953, did

---

[i] Sciama had suspected the factor 1, Schrödinger the factor 2, while the "Einstein-Radius" requires a factor 3. However, establishing the relation in general is the essential thing.

not know Einstein's article from 1911 and Dicke did not know either approach.

ON THE ORIGIN OF INERTIA

D. W. Sciama

(Received 1952 August 20)*

Summary

As Einstein has pointed out, general relativity does not account satisfactorily for the inertial properties of matter, so that an adequate theory of inertia is still lacking. This paper describes a theory of gravitation which

Title page of Sciama's original publication from 1953.

Sciama's idea however suggests yet another, slightly different, representation of the gravitational that we will explain in more detail at the end of the chapter. If instead of Newton's gravitational constant G, one considers the "Einstein constant" $\kappa = 8\pi G/c^4$ (with the unit $1/N$) referred to in chapter 6, then the result is not

$$4G \sum \frac{m_i}{r_i} \approx c^2$$

but simply $c^2 = \frac{2\pi}{\kappa \sum \frac{m_i}{r_i}}$ or even $\kappa \sum \frac{m_i}{r_i} = \frac{2\pi}{c^2}$. The key point of this reformulation is that the decrease of the speed of light is realized directly in the vicinity of masses. For this reason, κ would become, unlike G, a pure conversion factor of units. It offers the possibility of defining mass by means of inertia (with the unit s^2/m). We will come back to this in chapter 11.

NOT EXACTLY NEWS

Perhaps I should explain in a little more detail why I am so enthusiastic about Sciama's formula. We use to like ideas particularly, if they contain a seed that has already been created by our own mind. The origin of fundamental constants is something that I have pondered for quite a long time and since a con-

Chapter 8: Robert Dicke's simple completion

ference on the gravitational constant in Pisa (Tuscany) in 2002, G has bothered me particularly.

I can reflect best on physics during hiking, at a slow pace that leaves time for occasional short calculations. I remember that in June 2003 in beautiful northern Italy, I mused about the electrodynamic constants α and e for a full day, though without any result. Slightly frustrated, the next day I decided to tackle the problems of gravity. While climbing on a secluded path above Lake Idro, I decided to throw all my prejudices overboard and start over again.

The units $\frac{m^3}{s^2\,kg}$ of the gravitational constant G obviously required that the speed of light c^2 occurred therein, and I tried to follow this hypothesis. What remained when I took out m^2/s^2? The resulting meter per kilogram defied any reasonable interpretation. If I wanted to apply Mach's principle, all masses m_i would somehow have to be put into a formula containing m/kg. I wrote the sum $\sum \frac{r_i}{m_i}$ in my notebook, but I immediately noticed that it made no sense – larger masses would then carry less weight. Vexedly, I crossed out the formula, and then it dawned on me: one could also take the reciprocal sum $1/\sum \frac{m_i}{r_i}$. It represents the gravitational potential of the universe – but with the factor of G missing. That meant that the gravitational potential of the universe would be nothing other than the square of the speed of light!

I found this idea overwhelmingly attractive but at the same time it was obvious to me that such a simple formula must have been published somewhere already. Impatiently, I cut the hiking holiday short, and on the way back, – availing myself of internet access at the University of Innsbruck in Austria, I Googled Sciama's article from 1953. A little later, I also found Dicke's 1957 paper. Relating G and c was no fantasy! But I couldn't understand why these articles are so rarely quoted. There is a flood of publications on the cosmological standard model.

THE CLASSICAL TESTS

Let us return to our main theme. Is Dicke's version of 1957 really equivalent to Einstein's general relativity? To evaluate Dicke's discovery, let's look at what the agreement of general relativity with observation consists of.

There are four so-called classical tests of the theory, called light defection, gravitational redshift, radar echo delay and the perihelion advance of the planet Mercury. We already know the first test, the deflection of the star light by the tiny amount of about 1.7 arc seconds, observed for the first time in 1919. The effect has been confirmed on numerous occasions and has now reached a precision of 0.1 per cent.[47] Einstein's idea of 1911 correctly describes the result, if one uses Dicke's index of refraction $\alpha^2 = 1 + \frac{2GM}{rc^2}$.

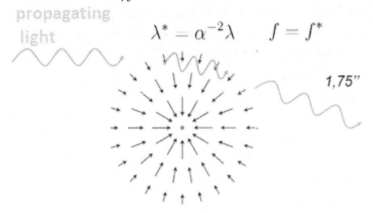

Schematic description of sunlight deflection. The speed of light c is reduced in the vicinity of the Sun by a factor $\alpha^2 = 1 + \frac{2GM}{rc^2}$ (M mass of the Sun, r distance from its center, G gravitational constant). As in conventional optics, light is therefore deflected towards lower c.

The gravitational redshift, instead, can be understood in conventional terms as loss of energy of a photon leaving a gravita-

Chapter 8: Robert Dicke's simple completion

tional field, whose color thus is shifted to the red.[i] The effect was shown first in 1969 in a spectacular experiment on Earth[48] and was also observed in 1972 in sunlight.[49]

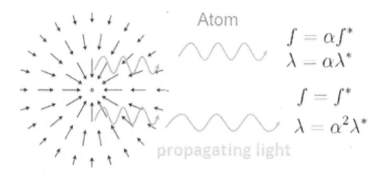

Illustration of the gravitational redshift. Atoms outside the gravitational field have wavelengths and frequencies enlarged by a factor $\alpha = 1 + \frac{2GM}{rc^2}$ each. On the other hand, light leaving the gravitational field must maintain its frequency f, so that the total change in the speed of light c, owed to the relationship $c = \lambda \cdot f$, must be contained in the wavelength λ. λ of the outgoing light is, thus, bigger by a further factor than the wavelength of the surrounding atoms. It appears, therefore, red-shifted.

In Dicke's theory of the variable speed of light, the effect is easily understood. In the gravitational field not only the frequencies f but also the wavelengths λ are lower, if the above symbolic notation $c\downarrow\downarrow f \downarrow \lambda\downarrow$ is used. c is lowered in a twofold manner. However, light that propagates in a gravitational field, despite the change of c, does not change its frequency, just as it does not do so in conventional optics. A source of light emitting continuous waves would otherwise "lose the beat," and the wave form would be destroyed by gaps. Thus, when light escapes from a gravitational field with the twofold lower c and its frequency cannot change, it needs to compensate by changing

---

[i] Because E=hf=hc/λ, λ increases.

the wavelength λ only by the twofold amount. This doubly shortened wavelength is then even longer than the wavelengths of atoms outside the gravitational field. Therefore, it is perceived as red-shifted.

While these effects were already known when general relativity was developed, the American physicist Irwin Shapiro in 1964 proposed a new and amazingly simple test. Light passing celestial bodies is not only deflected, but also delayed. Einstein could not foresee that such a measurement would ever become possible; however, radar technology developed in the 1960s (radar being an electromagnetic wave such as light) made it possible to send signals from Earth to Venus and to detect the reflected signal. The waves passing close to the Sun showed a delay of approximately 200 microseconds, as predicted by general relativity. In the variable speed of light formulation of general relativity this is easy to understand: the waves have indeed passed through an area with a lower c and consequently, are delayed.

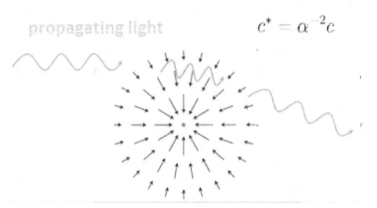

Schematic picture of the radar echo delay which is particularly easily explained by the variable speed of light. Since a ray of light passes through an area with reduced c, it arrives later.

Chapter 8: Robert Dicke's simple completion

SEEMINGLY A CLOSE MISS

These results were obtained quite naturally by Dicke,[i] unlike the case of the perihelion advance, a tiny shift of the elliptical orbit of planet Mercury, known since 1859. As it is not about light, but about a material body, this fourth test of general relativity is qualitatively different from the others and more difficult to derive. Dicke does not specifically say it in his article in 1957, but he had possibly failed to obtain the correct result.

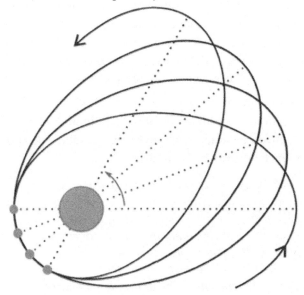

Illustration of the perihelion advance of the planet Mercury. The elliptical orbit shifts its position infinitesimally after each rotation. While the major part of this shift is owed to the influence of other planets, a residual 43 arc seconds per century is explained by general relativity only.

There is only an indirect reference to this in an article by another author,[50] who claimed that Dicke told him that the perihe-

---

[i] The radar echo delay, to be precise, which was explained afterwards.

Einstein's Lost Key

lion advance did not come out correctly – only 5/6 of the observed value. It appears however that Dicke had miscalculated, as did on another occasion we will discuss later on. Subsequent publications cite the correct perihelion shift with variable speed of light as well.[51]

A particularly comprehensible presentation (though in German) deserves to be mentioned here. In 1960, an article appeared in the *Annals of Physics* by the physicists Dehnen, Hönl and Westphal, Dehnen being a former chair of theoretical physics at the University of Constance. Under the title "A Heuristic Approach to General Relativity," the authors (without knowing about Dicke's article) considered a series of physical quantities that changed in accordance with a variable speed of light.

Before we go into details, a word about why this article hardly drew any attention. Instead of suitably highlighting the importance of their original approach involving variable speed of light, the authors belittled it in their introduction:

> *"It must be expressly stated that – and this actually goes without saying – Einstein's approach in setting up general covariant field equations and the search for their solutions are of course far superior to the method presented here. This is already clear from the fact our method can only describe physical effects of first and second order, at best."*

Yet the article does no less than explain all known tests of the theory with variable speed of light! What exactly is the "far superior" feature of the geometrical method, the reader may ask. To this day, no tests have been developed which are more precise than first or second order. One might think that the introductory remarks are a becoming sign of modesty, but perhaps they also demonstrate anticipatory obedience to the prevailing view in theoretical physics whereby one does not even appreciate one's own discovery. It seems that by 1960 it was already

Chapter 8: Robert Dicke's simple completion

unseemly for established scientists to question general relativity and its claim to absoluteness. That they had picked up one of Einstein's ideas was unknown to Dehnen and his co-authors!

A WORLD WITHOUT FIXED RODS

Because it is important in a cosmological context, let us consider the "physics of variable measuring rods" which was independently developed by Dicke and the German researchers. It has been mentioned that because $c = \lambda \cdot f$, the variability of the speed of light c influences both the frequencies f and the wavelengths λ of atoms (again, the factor $1 + \frac{GM}{rc^2}$ is denoted as α).

Therefore, all time and length measurements are automatically affected. One can imagine that every physical unit such as meters and seconds is subject to change. All speeds, for example, are measured in relation to the speed of light, or by the local units meters and seconds, if you like. Thus *all* speeds in a gravitational field are reduced in a twofold manner (because frequencies f and wavelengths λ contribute to it.)[i] The same principle applies to accelerations, which are reduced threefold because of the unit m/s^2. Since Newton's law states $a=F/m$, mass is inversely proportional to acceleration, and thus masses in a gravitational field are apparently *increasing*, namely by a threefold factor α^3, which was determined by both Dehnen and Dicke. To visualize this, imagine an environment in which all processes run more slowly: the inertia of masses seems to have increased.[ii]

---

[i] We consider the so-called first approximation in relatively weak gravitational fields, which however is sufficient for all astronomical data. Concretely, this means that the factor α does not differ substantially from one, thus approximately $1,01 \cdot 1,01 \approx 1,02$ or $1/1,03 \approx 0,97$ and so on. In general, $\alpha^n - 1 = n(\alpha - 1)$ holds.

[ii] Of course, such a change of measuring rods can only be perceived indirectly.

Einstein's Lost Key

Practically all physical quantities are subject to this variability, for every formula in physics has to remain valid, like the famous $\varepsilon_0\,\mu_0 = 1/c^2$ in electrodynamics. These electromagnetic constants, including the elementary charge, also change, but we will restrict our discussion to the mechanical quantities.

Quantity		Unit	Factor	Example	
speed of light	c	m/s	α^{-2}	0.98	↓↓
wavelength	λ	m	α^{-1}	0.99	↓
frequency	f	1/s	α^{-1}	0.99	↓
speed	v	m/s	α^{-2}	0.98	↓↓
acceleration	a	m/s²	α^{-3}	0.97	↓↓↓
inertial mass	m	kg	α^{3}	1.03	↑↑↑

Overview of the spatial dependencies of quantities in a gravitational field. Each physical unit such as meters or seconds is subject to change, and this affects all other quantities. In our example, α= 1.01, the factors in the fourth column show the corresponding change of each quantity.

ANY QUESTIONS?

Starting from the increased inertia of masses in a gravitational field, Dehnen and co-authors deducted a perihelion shift of the same amount as in general relativity, in agreement with all observations. The problem is actually solved. However, most physicists are unfamiliar with this approach, or at least do not believe in its completeness – it looks too simple.

The fact that the speed of light varies locally has entered conventional textbooks, even if it is usually referred to as "local coordinate speed without physical significance," The Belgian physicist Jan Broekaert published an excellent article,[52] in which he gave a general proof of the equality of the variable speed of light approach to general relativity. Broekaert collected numerous quotes from textbooks on general relativity that men-

Chapter 8: Robert Dicke's simple completion

tion variable speed of light (albeit in a misleading manner). The only thing missing is the simple statement: Gravity is nothing else than variable speed of light.

> Der Vergleich von (5.15c) mit (5.6c) und der nachfolgenden Rechnung zeigt völlige Übereinstimmung der Ergebnisse (bis auf den unwesentlichen Wert der Konstanten k); damit erhalten wir für die Perihelbewegung in der hier angestrebten Näherung
> $$\delta = 6\pi \frac{G^2 M^2}{a F''^2}. \qquad (5.16)$$

After a series of calculations, Dehnen and co-authors derived the famous formula for the perihelion shift (p. 396). "Comparison of (5.15c) with (5.5c) and the subsequent calculation shows a complete agreement of the results (besides the irrelevant constant k); therefore, we obtain for the perihelion advance in the desired approximation..."

By the way, the model of variable measuring rods nicely illustrates the origin of potential and kinetic energy, two well-known classical concepts. Kinetic energy is, according to special relativity, the increase of mass in the formula $E=mc^2$. Potential energy is explained quite simply by the variable speed of light[i]: the increase of c^2 in the formula $E=mc^2$.

In summary, one might be tempted to close the case and allow Dicke to take the credit owed to him. However, if mass really has an influence on the speed of light, then this would have another incredible consequence. Since the Big Bang, in every instant new light signals of previously unknown masses reach us, and consequently the speed of light would not only depend on position, as Einstein emphasized, but would also have to change over *time*. In view of new masses, c has to decrease. This sensational insight is Dicke's alone, and we'll take a look at the revolutionary consequences for cosmology that may arise in the two chapters that follow.

---

[i] Quantitatively, a subtlety remains here. Only a fraction of ¼ contributes to the potential energy (because ¾ is the inherent mass increase owed to inertia), but conceptually this does not matter.

Einstein's Lost Key

To the reader who is interested in further detail, another formulation is presented here which directly yields the dependence of the light speed from the masses. In turn, an expression for the gravitational constant G is derived, which was given by Sciama. Since it is possible to rephrase these calculations using some well-known theorems of vector analysis, it will become clear that the variable light speed approach is also very similar to Einstein's equations from a formal point of view.

Since these remarks require significantly higher mathematical knowledge than needed to now, the reader can, without compromising further understanding, immediately jump to chapter 9. There, another great thinker – Paul Dirac – enters the game, whose works are related to Einstein's idea – though this is unknown until today.

FROM THE NEWTONIAN LAW TO EINSTEIN'S EQUATIONS:

DERIVATION OF THE NEWTONIAN LAW FROM THE SCIAMA-DICKE FORMULA FOR THE SPEED OF LIGHT.

If we enter these technicalities, then Newton's law of gravitation follows from the above formula $c^2 = \frac{1}{\kappa \sum \frac{m_i}{r_i}}$. Because the gravitational potential, according to Sciama, is $\varphi = \frac{1}{4} c^2$, the local acceleration is obtained via differentiation:[53]

$$g = -\nabla\varphi = -\nabla \tfrac{1}{4} c^2 = -\nabla \frac{\pi}{2\kappa \sum \frac{m_i}{r_i}}.$$

(the operator ∇ "Nabla" means spatial derivative)

By applying the chain rule of differentiation we get

$$g = \frac{\pi}{2\kappa(\sum \frac{m_i}{r_i})^2} \sum \frac{m_i}{r_i^2} = \frac{c^2}{4 \sum \frac{m_i}{r_i}} \sum \frac{m_i}{r_i^2} = G \sum \frac{m_i}{r_i^2},$$

Chapter 8: Robert Dicke's simple completion

and suddenly Newton's inverse-square law emerges. In the second step the term $\frac{2\pi}{\kappa \sum \frac{m_i}{r_i}}$ was substituted by c^2, and finally, $\frac{c^2}{4 \sum \frac{m_i}{r_i}}$ was identified with the gravitational constant G, as Sciama and Dicke proposed.

Yet this calculation does not have much formal similarity with Einstein's equations. It can be shown, however, that with some transformations from vector analysis, the relationship $c^2 = \frac{2\pi}{\kappa \sum \frac{m_i}{r_i}}$ can be rewritten in a way that is reminiscent of Einstein's equations.[54] This was the original idea of Jan Preuss, a former student of mine, and it leads directly to the simple and suggestive form

$$-c^2 \Delta \frac{1}{c^2} = w,$$

in which the Laplace operator Δ appears, a differential operator mentioned in chapter 6. This operator is a close cousin of the Einstein tensor, if one takes curvature into account when taking the derivatives (so-called covariant derivative).

On the right-hand side we have w, as in the conventional formulation of the theory of relativity, a spatial curvature with the unit $1/m^2$. It came about as a result of two spatial differentiations, as described by the Laplace operator. In this respect, variable light speed contains a geometrized form of gravitation as well.

I hope I have not discouraged you by this technical aside. You can delve further into these matters, should you be interested.

Einstein's Lost Key

Chapter 9

The Genius Who Didn't Talk to Einstein

How Dirac's Large Number Hypothesis enters the game

The history of science is not a chain of events, but rather a network of interwoven developments which are difficult to grasp simultaneously. In chapter 10 the full extent of Einstein's idea and Dicke's brilliant extension of it become visible, but before then another great thinker joins these two, whose ideas combine with variable speed of light in a surprising manner: Paul Dirac.

The silent Englishman, as much a loner as Einstein, played a prominent role as a founding father of quantum mechanics. In 1926, two rival forms of the new theory of the atom existed: Werner Heisenberg's matrix mechanics and Erwin Schrödinger's wave equation. Dirac was able to show that they were equivalent formulations, an achievement that could probably find due appreciation only when the two versions faced each other as irreconcilable formalisms. Furthermore, Dirac also developed an equation that embedded Einstein's special relativity into quantum mechanics, describing a puzzling characteristic of the electron, the 'spin'. In 1933, together with Schrödinger, he received the Nobel Prize for his development of the Dirac equation.

THE NUMBER WHISPERER

His greatest accomplishment, however, is something else. The idea has not been completely forgotten, yet its consequences have been greatly underestimated since then. It concerns a connection between cosmology and particle physics which is known as *Large Number Hypothesis*. Unlike many modern fantasies of unification, Dirac's consideration relies on a concrete observation. He was particularly interested in combinations of fundamental constants that yielded pure numbers – a profound riddle of Nature when we measure certain physical quantities.

Paul Adrien Maurice Dirac (1902-1984)

Chapter 9: The genius who didn't talk with Einstein

For instance, there is the mass ratio of protons and electrons that amounts to 1836.15..., a number that still awaits an explanation today.[i] It was this problem that drove Dirac to derive his equation. Evidently, he himself viewed his Nobel Prize as a side issue and never stopped stressing the fundamental importance of such questions.[55] Likewise, Dirac pondered on the mysterious fine structure constant *1/137.035999*. Young physicists seeking his advice but who failed to show a proper awareness of the problem were dismissed with the words: "Come back when you have worked out this number!"

Here, we are dealing with an even stranger number. Edwin Hubble's discovery of the cosmic redshift, which led to a first estimate of the size of the universe, was a turning point in Dirac's scientific endeavors. In the 1930s, he started to think about the biggest structures in the universe, and this led him to the large number hypothesis. At the same time Hubble, by dividing the distance of galaxies by their receding velocity, calculated a time scale that marked the approximate age of the universe. As a ballpark figure, the greatest distance in the cosmos is therefore the corresponding number in light years, if one imagines light propagating all the time since the Big Bang.[ii]

On the other hand, Dirac had pondered for many years the question of why the electric force in the universe is so much stronger than the gravitational force, despite the fact that the laws ruling these forces are so similar in structure: Newton's inverse square law $F = \frac{GMm}{r^2}$ is reflected in Coulomb's law of electrostatic interaction, except that Coulomb's force is generated by charges Q and q, not masses: $F = \frac{Qq}{4\pi\varepsilon_0 r^2}$. Unlike gravita-

---

[i] If one ignores the ridiculous claim that the 'Higgs field' is responsible for mass.
[ii] The cosmological standard model makes further assumptions which are of no interest in terms of this approximate consideration.

tion (determined by G), the intensity of the force is determined by the value of the constant $\frac{1}{4\pi\varepsilon_0}$. If we consider a hydrogen atom in which both forces are at work when a proton and an electron (with masses m_e and m_p) orbit one another, how big is the ratio of the two forces? Putting in the figures (the distance r cancels out), we get the value $\frac{F_e}{F_G} = \frac{e^2}{4\pi\varepsilon_0 G\, m_p m_e} = 2.27 \cdot 10^{39}$, an amazingly huge number with almost forty digits.

Nature seems to have gone mad by setting such a disproportionate ratio for the two dominating forces, dashing our hope of understanding structures that are ultimately similar. How, wondered Dirac, could such a surreal value ever be derived by pure mathematics?

COINCIDENCE

It must be borne in mind that Dirac, like Einstein, was firmly convinced that seemingly inexplicable numbers were not God-given but had a meaning, and he believed that deciphering Nature's writings was the mission of a physicist. If one browses through any biography of Dirac, it is immediately clear how much he would despise the attitude of contemporary physicists who deny problems like this.

Albeit Dirac thought it hopeless to calculate such large numbers by mathematics, he suggested there might be another occasion on which nature reveals such a gigantic ratio. Edwin Hubble's measurements in the 1930s must have been a revelation to him. Yes, it is the size of the cosmos! Dirac compared the extension of the universe as measured by Hubble with the size of the proton. Since Rutherford's experiments in 1914, this was known to be about 10^{-15} meters, and the ratio of the sizes of the universe and the proton turned out to be that huge number with forty zeros. Nowhere else in nature are such huge numbers found, and here they match!

Chapter 9: The genius who didn't talk with Einstein

Of course, these numbers are not exactly the same, but this is of minor importance. All astronomical measurements contain huge inaccuracies, and many quantities can only be estimated by orders of magnitude. Above all, however, such a coincidence alone cannot provide an exact mathematical relation; a complete theory needs to be developed. As Dirac noted in his article[56] "A New Basis for Cosmology" in 1938, various other factors could be contained in these numbers: the fine structure constant (about 1/137), or the factor 1836.15, if one considered two electrons instead of the hydrogen atom (which raises the number 10^{39} to 10^{42}). All this should not distract us from the fact that the coincidence is, as Dirac wrote, "highly remarkable."

> A new basis for cosmology
>
> By P. A. M. Dirac, F.R.S.
>
> St John's College, Cambridge
>
> (Received 29 December 1937)
>
> 1. Introduction
>
> The modern study of cosmology is dominated by Hubble's observations of a shift to the red in the spectra of the spiral nebulae—the farthest parts of the universe—indicating that they are receding from us with velocities proportional to their distances from us. These observations show us, in the

Front page of Dirac's original publication.

Why is this observation assessed in such a different manner by today's physicists? Since the appearance of two similar numbers in physics is in itself nothing spectacular, this "accident" is often presented in a derogatory way. However, the value of Dirac's hypothesis depends on whether one sees a problem in unexplained numbers in nature or not. Einstein, Dirac and others in their time found the idea of inexplicable numbers distasteful, whereas modern physics has got used to an abundance of such arbitrary parameters. Correspondingly, in today's research environment, there is no incentive at all for wondering about such coincidences, whereas Einstein and Dirac considered the occurrence of arbitrary constants in physics a priori unlikely. Obviously, opposing philosophies clash here – I stick to that of Einstein and Dirac.

> *People are entirely too disbelieving of coincidence – Isaac Asimov*

ACCIDENTAL? – NO WAY!

The coincidences observed by Dirac are however much more pronounced than is commonly known. A waiter failing to issue a customer's change is something that can happen occasionally, but if it happens again with the following customer, it is unlikely to be an accident. Dirac, when pondering the size of the universe, discovered another, independent, relation, which was no less mysterious than the first. Hubble had determined the expansion of the universe, and thus one could also estimate the matter contained in it. Dirac now wondered how many particles there were in the universe. He divided Hubble's mass estimate by the mass of the proton[i] and got about 10^{78}, a number with almost eighty digits! Dirac was delighted. The number of particles was obviously the square of that number 10^{39}, since the order of magnitude is doubled when squared. The first coincidence regarding size had been noted also by Eddington, but it was Dirac alone who discovered this mystery relating to masses. From a physical point of view, the second coincidence sounds even more incredible, but the repetition strongly suggests that there is a reason behind it.

If you were already reflecting about constants of nature, perhaps you have noticed a consequence of Dirac's hypothesis. According to quantum mechanics, to each particle a wavelength that depends on mass may be assigned. If one transforms the energy of this mass by using Einstein's famous formula $E=mc^2$

---

[i] The proton is the most basic atomic nucleus (of hydrogen); thus Dirac may also have considered the number of atoms in the universe.

Chapter 9: The genius who didn't talk with Einstein

into the energy of a photon according[i] to $E = h \cdot f$, the result is $\lambda = \frac{h}{cm}$, which is called the Compton wavelength.

The greater the mass of a particle, the smaller its quantum mechanical wavelength; therefore, the wavelength normally has nothing to do with the actual size of the particle. Only in the case of the proton do they agree approximately. Thus one can conclude that the proton plays a prominent, fundamental, role in nature – a point of view all reasonable people held before the invention of the 'quark' model. According to today's philosophy, instead, the proton's size and weight are not just random quirks of nature. However, the relationship $h = c\, m_p\, r_p$ (mass and radius of the proton) tells another story, and in a few steps,[57] it can be derived from Dirac's hypotheses, if one combines it with the 'flatness' $G \approx c^2 \frac{R_u}{M_u}$.

By the way, every now and then Dirac's coincidences are newly 'discovered' and published by researchers who are unaware of Dirac's paper of 1938. This happens because the agreement in orders of magnitude is still there even if one includes (as Dirac had already suggested) factors such as the fine structure constant $\frac{e^2}{2hc\varepsilon_0} = 1/137$, or the proton-electron mass ratio 1836.15.... Carl Friedrich von Weizsäcker, a student of Heisenberg, proposed, for example, the formula $m_p^3 \approx \frac{h^2}{G R_u}$ which is a rather crude match. Helmut Söllinger, an engineer from Vienna, recently discovered a similar coincidence, which is related to Dirac's idea but exact within a few per cent:[58]

$$\sqrt{m_p m_e} \approx \sqrt[3]{\frac{e^2 h}{4\pi\varepsilon_0 c G R_u}}$$

---

[i] The discovery of the fundamental significance of this formula we owe to Einstein as well. $c = \lambda \cdot f$ always holds.

After all, this term also contains the electrical constants of nature. Such thoroughly interesting speculations are however so far from the mainstream that 'established' physics journals would reject them.

Why are coincidences such as Dirac's considered exotic? Assuming that the number of hydrogen atoms in the universe is proportional to the square of its size indeed appears grotesque: as if the amount of matter in the universe had to do with its surface, rather than with its volume. Even today, this seemingly contradictory fact makes it hard for many to appreciate Dirac's hypothesis. Needless to say, the standard cosmological model is irreconcilable with it; the entire concept of the expanding universe, as we understand it so far, is challenged. Thus for the adherents of the standard model it is comfortable to stress that there isn't yet any theory which can predict Dirac's coincidences accurately. In this context, Sir Arthur Eddington once recommended sarcastically, "not to put overmuch confidence in the observational results ... until they are confirmed by theory."

To round off the value of Dirac's observation, however, one should mention that it is in complete harmony with Ernst Mach's thoughts on gravity, though Dirac apparently never dealt with Mach.[59] But probably he was convinced as well that that the relation $\frac{c^2}{G} \approx \frac{M_u}{R_u}$ had a meaning. The fact that Dirac considered the size and the mass of the universe, the two quantities that Mach also related to the origin of gravity, constitutes another piece in this fascinating puzzle.

MACH'S PRINCIPLE 2.0

However, Dirac's observation goes beyond Mach's principle. Imagine that the number of particles in the universe was a billion times larger, while simultaneously their mass was a billion times smaller. This would change nothing about Mach's principle (or 'flatness') but it would alter Dirac's observation. In other

Chapter 9: The genius who didn't talk with Einstein

words, Dirac was the first to insinuate that the size and the mass of elementary particles had a meaning, and that it is no coincidence that they are as large and heavy as they are. Who thought soothe same? You've guessed it – Albert Einstein:[i]

> "The real laws of nature are much more restrictive than the ones we know. For instance would it not violate our known laws, if we found electrons of any size or iron of any specific weight. Nature however only realizes electrons of a particular size and iron of very specific weight."

One can be sure that Einstein would have been fascinated by the Large Number Hypothesis had he known it! Dirac would have paved the way for a possible unification of physics with his own observations that, as he said himself, allow one to assume "a deep connection between cosmology and atomic physics."

This coincidence has been woefully underestimated until the present day and I cannot, at this point, refrain from observing the contrast with the superficialities which today pass for fundamental physics. Everyone who is familiar with the basics of theoretical physics knows that its two main pillars, quantum mechanics and general relativity, are incompatible. Admittedly, this lack of unification is the biggest problem, the failure to synthesize the grand with the small, the macrocosm with the microcosm. On the other hand, it should be obvious to any scientist that progress in science does not consist of massaging academic egos; rather it means comparing the quantitative predictions of a theory with experimental evidence.

Considering general relativity, i.e. gravity, in the most elementary quantum system, the hydrogen atom, yields the easily

---

[i] Needless to say, the standard model of elementary particle physics is incapable of making sense of these values.

measurable yet enigmatic number $2.27 \cdot 10^{39}$. It is therefore crystal clear that any theory that hopes to unify quantum theory with relativity must calculate this number and explain it, if it does not want to end up in futile verbiage.

Yet generations of theoretical physicists deal with such useless babble and, simultaneously, dare to defame the only viable idea (due to Dirac) as "numerology." If only Dirac could return to kick these acrobats of mathematical formalisms out the door by saying: "Return after having thought this number over!"

Perhaps you are startled by my tone, but I cannot conceal how much the ignorance of today's research regarding Dirac's numbers annoys me. In numerous books, the ratio of electric and gravitational force is mentioned only casually as a curiosity, as if Paul Dirac had never existed. Quite a long time after graduation and almost by chance, I came across the original article – which then left me speechless because of its intriguing content. I wish I had realized the significance of this discovery much earlier. Dirac's observation was also the reason why, after years of work, I turned away from the Einstein-Cartan theory of teleparallelism[60] of 1930. As enticing as their ingenious geometrical unification of electricity and gravity may appear on a formal level it certainly cannot explain the quantitative difference of the two forces.

POSSIBLE HASTE

As in the case of Mach's principle, Dirac's idea is poorly appreciated – though it is a premature judgement. Unfortunately, however, in his article from 1938, Dirac himself wrote things that rendered his theory a little vulnerable. This does not excuse the groupthink of scientists who believe that Dirac's ideas are outdated, without having seen a single page of the original article. But Dirac took a risk (understandable perhaps in view of his enthusiasm) and claimed that his hypothesis would force the gravitational constant to decrease with time.

Chapter 9: The genius who didn't talk with Einstein

He suspected a relation between the numbers

$$\frac{e^2}{4\pi\varepsilon_0 G\, m_p m_e} \approx 10^{40} \approx \frac{R_u}{r_p},$$

while the radius of the visible universe R_u apparently grew ever larger each day. Thus it was not too fanciful to conjecture a decrease of the gravitational constant G. Although a variety of observational tests have been dedicated to this question, the variability of the gravitational constant has so far not been proven.[61] There are articles claiming that a decrease by the amount predicted by Dirac is precluded, but many of them make over-optimistic assumptions about the accuracy of their data. We will come back to this highly interesting question in chapter 13.

Dirac's Large Number Hypothesis, which was touched only tentatively by his former colleagues, was forgotten over the years. He may even have moved away from it himself (from the second coincidence regarding mass). This was what Pascual Jordan claimed at least when, admiringly, he wrote in 1952:

> *"I consider Dirac's ideas for one of the greatest insights of our time; the further study of these ideas has to be one of our principal tasks."*

Unfortunately, after World War II, Jordan was still not cured of Nazi ideology, a fact that understandably compromised his reputation (his contributions to quantum mechanics were close to worthy of a Nobel prize). We will discuss Jordan's ideas later in chapter 13 in terms of his use of Dirac's numbers. Despite this support, Dirac's idea remained completely separated from mainstream cosmology, in a kind of forgotten niche. This is sad, because we are dealing with a deep puzzle which is of paramount importance for the very fact that it turns traditional convictions upside down.

> *An idea that is not dangerous is unworthy of being called an idea at all.* – Oscar Wilde

THE DYNAMICAL COSMOS

Besides the remaining experimental uncertainty there are also theoretical reasons why one should not dismiss Dirac's general idea even if the special form published in 1938 is not fully correct. Dirac examined the variability of several physical quantities including, among others, the gravitational constant. But he forgot something – that wouldn't have happened if he had talked to Einstein about his idea from 1911: variable speed of light. Dirac came so close to hitting the mark. In the last section of his publication (to which one finds hardly any references in secondary literature) he developed a rather visionary model of a dynamically evolving cosmos.

Dirac introduced an absolute time, which is a useful descriptive tool, to express the dynamic evolution of the size of the cosmos as a function of time. Thus Dirac moved away from the naive idea that the universe expands uniformly with time. To put it simply, this is the case if the barely visible edge at a distance R ("radius of the universe") recedes at the speed of light, mathematically expressed by a proportionality $R(t) \sim t$. Dirac suggested instead that the size of the universe is a function of time with a different exponent: $R(t) \sim t^{\frac{2}{3}}$. That would mean, for example, that the universe at a thousandfold age would only grow to a hundredfold size.[i] As the visible horizon (R) of the universe is defined as the maximum distance from which we can receive light signals, it actually follows that in Dirac's model the light speed can no longer be constant. However, he did not explicitly run with the idea.

Such a model implies that our entire perception of the universe relies on a frame of variable time and length scales, from which we can unveil the true dynamics only by indirect conclusions. Even though the relation $R(t) \sim t^{\frac{2}{3}}$ given by Dirac is not

---

[i] Recall that $100 = 10^2$ and $1000 = 10^3$.

Chapter 9: The genius who didn't talk with Einstein

exactly correct, the idea in general remains revolutionary. For very little was missing and Dirac would have realized the close parallel with Einstein's variable speed of light. It is tragic, and from today's perspective almost incredible, that the two greatest physicists of the twentieth century never talked to one another about cosmology.

epoch. The distance between neighbouring spiral nebulae, expressed in the new units, will vary with the epoch according to the law

$$f^*(t) = tf(t) \propto t^{\frac{1}{3}},$$

and hence $\quad f^*(t^*) \propto t^{*\frac{1}{4}}.$ (6)

Dirac's 1938 model in which he considered an expansion of the universe nonlinear in time.

SEPARATED GENIUSES

The only hint, albeit an indirect one, that Einstein had heard of similar ideas, is found in his correspondence with Ilse Rosenthal-Schneider in 1945. Rosenthal-Schneider had addressed Eddington's ideas on the "non-empirical constants." According to her, his "completely non-Kantian apriorism" was, with respect to laws of nature, "clearly unsustainable." Einstein responded to this in a letter dated April 23, 1949:

> "Eddington has made a series of spirited suggestions, which I, though, haven't followed up on. In general, I believe that he has been fairly uncritical of his own ideas."

Einstein expressed himself quite frankly about Eddington's conjectures, which indeed were not all well-founded. Once Eddington claimed to have derived the fine structure constant from pure logic, obtaining the number 136. His logic, however, was flexible enough to adapt to the new measuring value 137 shortly after. His colleagues then mocked him calling him "Mr Adding-One."

Einstein's Lost Key

It remains unclear whether Einstein's blunt critique was referring to that episode or to another claim regarding the number of particles in the universe. Eddington guessed that the actual number 10^{78} originated from $137 \cdot 2^{256}$, something that can be called, in fact, numerology. Quite clearly, however, Einstein knew nothing of Dirac's more elaborate thoughts, or he would have inevitably mentioned him in this context. In Princeton's library, Einstein certainly had access to the journals *Nature* and *Proceedings of the Royal Society*, but he was notorious for not being an avid reader of literature. The scientists who surrounded him in Princeton were concerned with entirely different matters. Thus Dirac's article probably remained unknown to Einstein until his death.

Perhaps it is easier to understand why Dirac did not know Einstein's article in *Annalen der Physik*[i] – he was a 9-year-old boy at the time. Maybe he briefly noticed it at a later date, but certainly he did not see the link to his own thoughts. It is more likely however that in 1938, when Dirac's interest in cosmology was triggered, the 27-year-old Einstein article was simply too far away for him to take note of it. He could not expect that anything relevant to cosmology had been written before Hubble's discovery in the 1930s.

In addition, Dirac was too much of an original thinker to feel the need to insert his ideas into an existing set of ideas, even Einstein's. His article of 1938 refers very little to general relativity and consequently the concept of the speed of light as such is not explored. Much later, in 1968, Dirac made a kind of general plea for the variability of constants of nature:[62]

> *"Theoretical workers have been busy constructing various models for the universe based on any assumptions that they fancy. These models are probably all wrong. It is usually assumed that the*

---

[i] Not to be confused with *Annals of Physics*.

Chapter 9: The genius who didn't talk with Einstein

> *laws of nature have always been the same as they are now. There is no justification for this. The laws may be changing, and in particular, quantities which are considered to be constants of nature may be varying with cosmological time. Such variations would completely upset the model makers."*

It is utterly astonishing that he excluded perhaps the most important constant of nature, c, from this possibility. Or had he already been influenced by the disastrous tendency in theoretical physics to view c as an irrelevant conversion factor between space and time? We do not know. It is a tragedy that the link to Einstein's revolutionary idea of 1911 has never been established, and it is even harder to understand why Dicke's theory of 1957 did not completely incorporate his own Large Number Hypothesis. Yet there is a marvelous link between the two approaches which I shall explain in the next chapter.

Einstein's Lost Key

Chapter 10

Big Bang Without Expansion

Einstein's desired cosmos

In this chapter, I shall show how all-encompassing the consequences of Einstein's revolutionary idea turn out to be. In Chapter 8, an explanation of the gravitational constant G appeared for the first time because Dicke succeeded in linking the Einstein formula of variable speed of light to the universe. Ultimately Dirac, who had also pondered on the gravitational constant, discovered a puzzling relation between elementary particles and the universe, revolting against all conventional wisdom.

Now it has become clear that these separate discoveries form one big picture. Once again it is Dicke to whom we owe the decisive contribution. Besides the correction of Einstein's omission and the brilliant inclusion of Mach's principle, Dicke had a third spectacular insight into the consequences of variable speed of light for the evolution of the cosmos which revolutionizes the popular picture of the Big Bang.

This pearl is hidden in an unusual place. It is to be found on the last two pages of Dicke's 1957 publication where, in a paragraph on cosmology, he breaks with the dogma persisting since 1930: the Hubble expansion of the universe. Basically, Dicke says that the interpretation of the red shift, which led to the picture of an expanding universe, is an illusion. A revolutionary thesis, a simple explanation: we are dealing with the direct consequence of pushing Einstein's formula of 1911 further.

Dicke showed that in the formula which reproduced the laws of gravity all masses have an effect on the local speed of light: c

is smaller the closer one comes to the mass, but also the more mass is present within the visible horizon. Obviously, at any instant signals can reach us from ever more distant celestial bodies and therefore the visible mass within the cosmic horizon steadily increases. Thus, the speed of light must decrease on cosmological timescales.

A HIGH-SPEED MOVIE OF THE COSMOS

This variability of c has consequences for all other constants of nature, and, naturally, for all measurement scales with which observations are made. Since we only receive this relative information about our physical environment it is not easy at all to deduce what is actually going on.

We saw in Chapters 7 and 8 that variable *space* scales (usually, wavelengths of atoms act as measuring rods) in a gravitational field produce general relativistic effects. Dicke realized that the speed of light and accordingly the scales also had to be variable in *time*. This is all to do with the formula $c=\lambda \cdot f$ that connects the speed of light with wavelengths and frequencies, i.e. to the spatial and temporal scales (the latter is actually $\tau = 1/f$).

As we saw in Chapter 8, the explanation of classical tests such as light deflection require that while adjusting its wavelength λ (spatial scale), light has to keep its timescale, i.e. its *frequency f* constant. This is similar to conventional optics where light refraction occurs with a shortening of the wavelength λ at a constant frequency f. Dicke now considered the reverse case and deduced that during its propagation across the cosmos light must keep its *spatial scale* λ while the decrease in c only shows up in variable timescales f. He expressed it by referring to the energy $E=h \cdot f$ of a quantum of light[i] (p. 374):

---

[i] Blue light has a shorter wavelength λ, and because of $f = c/\lambda$, a higher

Chapter 10: Einstein's desired cosmos

> *"The photon emitted in the past has more energy than its present counterpart. This might be thought to cause a 'blueshift'. However, the photon loses energy with increasing time at twice the rate of loss characteristic of an atom, and hence there is a net shift toward the red."*

Dicke supported his claim with a calculation involving Maxwell's equations.[i] How can we illustrate this rather abstract idea? Light that travels through cosmic distances (that is, billions of light years), feels the continuous decrease of the speed of light, while the *spatial* variation of c is negligible in such large distances. Now imagine an extended light ray, such as a laser beam trapped between mirrors that are many kilometers away from each other. Light under such conditions forms a standing wave, but when c decreases, light must *keep* its wavelength λ. Otherwise, a leap in the waveform would open up somewhere – impossible, because for symmetry reasons the wave would not know where to shorten itself first. During cosmic propagation, light maintains its wavelength. This deduction is essential.

When λ remains the same, because of the formula $c=\lambda \cdot f$ the entire change in the speed of light c has to occur in the frequency f. If we again use our suggestive notation, we can write it as $c \downarrow\downarrow f \downarrow\downarrow$. As stated above, light that travels over large distances in the universe does not alter its wavelength λ.

---

frequency f; thus it has more energy.

[i] Dicke writes in addition (p. 374): "Although the photon concepts were used to obtain the galactic redshift, these particle ideas are not necessary. It is easily seen from Maxwell's equations that for time-dependent, but space-independent [speed of light], an electromagnetic wave propagates without change in wavelength."

ONLY LIGHT SPREADS OUT, NOTHING ELSE

Let us remind ourselves however that in the case of local atoms the decrease of c is equally distributed on frequency f and wavelength λ ($c \downarrow\downarrow f \downarrow \lambda \downarrow$). Compared with the local atoms that have participated in the shortening of λ, the wavelength of the incoming light from the cosmos appears larger, i.e. shifted to the red. This is Hubble's observation! The redshift of the light of distant galaxies that has been interpreted since the 1930s as an expansion of the universe can therefore be simply explained by the Einstein-Dicke formula of variable speed of light, once we apply it to the cosmos.

We see more and more masses out in the universe, and due to their influence, the speed of light is decreasing. The motion of matter or an expansion of the universe in the traditional sense is therefore not necessary in this model. The cosmic horizon, i.e. the border from which light emitted by distant objects reaches us, simply extends because light cannot do anything else but spread out. Only in this limited sense does the term "expansion of the universe" make sense in Dicke's model. The matter contained in the universe instead stands still, apart from relatively slow local motion. Considered the other way round, the speed of light is simply a quantity that describes how fast the cosmic horizon expands.[i]

The radical simplicity of Dicke's model is remarkable; in conventional cosmology (which we will take a look at in Chapter 12) the speed of light and the expansion rate are two different things, an assumption that has led to considerable confusion. Dicke cut the Gordian knot: the speed of expansion *is* the speed of light.

---

[i] Mathematically, the increase of the horizon is described by $\frac{dR}{dt} = c$ or $\dot{R} = c$.

Chapter 10: Einstein's desired cosmos

What is usually measured as the rate of expansion of the universe, the so-called Hubble constant, thus simply reflects the age of the universe, and incidentally explains why the values in the conventional model match so well.

BIG BANG YES, BUT NO EXPANSION OF THE UNIVERSE

We now have to branch out a little in order to understand the complete picture of the variable cosmos Dicke envisioned. Since different physical quantities vary in a different manner, at some stage it becomes easier to express them quantitatively, i.e. with formulae (if you don't share this view, don't worry if you skip the formulae.) First of all, we have to describe the expansion of the universe on a large scale. If the speed of light c really slows down with time, it means the horizon R (or "Radius" of the universe) does not increase proportionally with time but more slowly. How does one express the correct time dependence in a precise manner?

Technically speaking, one has to solve a so-called differential equation. From what has been said it is easy to see that the solution must have a certain form. According to the formula developed in Chapter 8

$$c^2 = \frac{2\pi}{\kappa \sum \frac{m_i}{r_i}},$$

the square of the speed of light is inversely proportional to the sum $\sum \frac{m_i}{r_i}$ that occurs in the gravitational potential. The average distances r_i of these masses will however grow in the same manner as the visible horizon R. Then, the number of the particles contained in the sum is proportional to the volume of a sphere with radius R, therefore $V \sim R^3$. If we put these dependencies in the above formula, we end up with $c^2 \sim \frac{1}{R^2}$. As c is nothing else but the growth of the radius R over time then[i] it follows that $R(t)$ must be proportional to the square root of time t: $R(t) \sim t^{\frac{1}{2}}$, in interesting proximity to the proposal by Dirac, who gave an exponent $\frac{2}{3}$ of time.[63]

We must, however, define the concept of time very carefully here. Because atoms, from which all clocks are built[ii], change their frequencies and thus their timescales, one must take into account how the perceived time is distinguished from the "real" time. The formula $c = \lambda \cdot f$ appears again, showing that the variability of c equally distributes[iii] to the frequencies f and the wavelengths λ.

In fact, for a mathematical model it makes sense to introduce an absolute, uniformly running time t. Subsequently, this absolute time t can be compared with the perceived time t' that, though distorted, is the basis of observations. All this was mentioned by Dicke, and the system of varying timescales is completely analogous to the changing spatial scales which we spoke about in Chapter 8.

However, the term "absolute" time sets alarm bells ringing for many physicists, who have eagerly learned from Einstein's theory that time is always "relative." But Einstein had this insight precisely because he focused on quantities that were measurable[iv] – exactly that what we are doing here. To describe the varying scales correctly, we therefore choose an absolute space-time system, which is, I repeat, not directly observable. So to speak, absolute time is good for relativity. The method used here in no way contradicts the principles of relativity.

---

[i] The time derivative reduces the exponent of t by one.
[ii] Of course, it does not matter whether it is conventional clocks or modern atomic clocks, for obvious reasons.
[iii] To be precise, this is necessary to fit the tests of general relativity.
[iv] A different kind of variability obtained in the special theory of relativity.

Chapter 10: Einstein's desired cosmos

ILLUSIONS ABOUT THE PASSAGE OF TIME

Timescales, which are absolute on the one hand and variable/observable on the other, help to avoid a series of weird consequences stemming from the assumption that the universe began at a distinct point in time, commonly denoted as the Big Bang.

If we define the time shortly after the "beginning" of the universe $t=1$ as the moment at which the universe was the size of an elementary particle, and set all other variables at that moment to $\lambda=1$, $f=1$, $R=1$, $c=1$ etc., then the evolution of time described above can be visualized as follows. At time $t=10,000$ the universe had grown only to $R=100$ (the square root of $10,000$), whereas the speed of light had fallen to $1/100$ of its initial value. Because $c = \lambda \cdot f$ the wavelengths λ and frequencies f declined to $1/10$ of their initial value, while the timescales τ ("tau," which are inversely proportional to the frequency) thus have the duration $\tau=10$. Once we realize that all time measurements involving clocks are performed with such timescales, we obtain an entirely new picture of the origin of the universe.

The conventional idea that the universe just came into existence at $t=0$ is in any case fairly naive, and in terms of natural philosophy supremely unsatisfying. In this new model of variable scales one can, of course, also set absolute time to $t=0$. But as one approaches $t=0$ the frequencies sharply increase, and the time steps τ are correspondingly small. As a result, the perceived time measured as multiples of these small time steps τ grows larger and larger, finally – at $t=0$ – becoming infinitely large. Accordingly, the universe has an (unobservable) beginning, but it had apparently an infinite number of time steps – that we might imagine as seconds – in the past.

The concept of the age of the universe thus becomes truly relative, not only because seconds pass but also because the duration of a second changes. During cosmic evolution, the time steps in which that age is measured become larger and larger. In the relatively long time steps that apply today, the beginning of the universe appears to be only a finite period in the past.[i] This beginning could be called the Big Bang even if the situation were completely different from conventional cosmology. Here, too, an old puzzle can be solved from an unexpected perspective.

DIRAC'S NUMBER APPEARS

Let us summarize this astonishing picture of variable scales once more: If 10,000 time steps since the Big Bang have passed, that beginning appears to be only 1,000 time steps in the past because those steps are now 10 times as long as they used to be. Conversely, the length scale has shortened to *1/10*, so the length that has an actual value of 100 appears to be 1000. So we conclude that extension takes place in proportion to time (1000 length units in 1000 time steps), because we cannot directly perceive the variability of the speed of light (indirectly, of course, we can deduce it from the cosmological redshift). In our model example, then, the initial volume would have grown by a factor of 1.000.000: the size of 100 raised to the third power. But since we still measure 1000 as a size factor, 1.000.000 would look like the size to the second power – a most important point, which we shall examine more closely.

We could reflect at this point on how the gravitational constant G changes the term that according to Sciama represents G: it really ought to decline, as predicted by Dirac. It is not at all clear, nonetheless, that this decline would become visible in any experiments designed for that purpose. Measurements of the

---

[i] In Dicke's model of variable time stages this age would be exactly a quarter of the common interpretation of the measurement, i.e. about 3.5 billion years.

Chapter 10: Einstein's desired cosmos

distance to Mars, for example, which are regarded as proof of the immutability of G, could well be affected by variable scales. We shall come back to this in Chapter 13.

If we extend the foregoing example of the 10.000 time steps to the current radius of the visible universe, viz. 10^{39} proton radii (we can define this as one), then we would actually already be living in the epoch $t=10^{52}$. Since the time steps have extended by a factor of 10^{13} since then, we get the impression of $t=10^{39}$. Correspondingly, the size of the universe is only 10^{26} (the root of 10^{52}), although, via the shortening length scales, it looks like 10^{39}. Now it gets really interesting: consider the volume of the enclosed space in the horizon, with the natural assumption that the density of elementary particles is the same everywhere in absolute space.

Since the spherical volume is proportional to the third power of the radius R, the volumetric size of the universe is in the order of 10^{78} – and contains about 10^{78} particles. That is Dirac's observation! As 10^{78} is the square of the number 10^{39} it is, in this model, perfectly intelligible that the number N of particles in the universe is proportional to the square of its extension. What previously seemed mysterious and downright illogical becomes inevitable in Dicke's model. It is amazing that Dicke's idea solves this seemingly unfathomable mystery, and of course Dirac's hypotheses are even more convincing thanks to this surprising connection with general relativity.

There was only one problem: Dicke had worked out the ingredients for this conclusion but, just before the end, had miscalculated in a relatively trivial way. He overlooked at one point (p.375) that the wavelength of atoms, whose variability he had previously given correctly (p. 366), influenced the length measurements. Dicke correctly gave the number of particles as proportional $t^{\frac{3}{2}}$ but forgot that in the perceived time $t^{\frac{3}{2}}$ looks like t'^2. In a subsequent unfortunate paragraph he justified this wrong result. The deviation from observation was obvious, but he tried various physical explanations: for example, that the universe was "relatively young." Instead, he could have brilliantly verified Dirac's hypotheses.

> $\tau^3 \sim \epsilon^3$ and the average distance between particles
> as $\epsilon^{\frac{1}{3}}$. This gives a total number of particles which
> varies as $\tau^{\frac{3}{2}}$ instead of τ^2 as suggested by Dirac's
> considerations.

Dicke's erroneous conclusion regarding absolute and observed time

DICKE'S GREATEST BLUNDER

It is incredible how Dicke missed the answer within his grasp. Perhaps he would have seen it if he had put in concrete numbers, instead of taking only the abstract exponents into consideration. Apparently, in the disagreement with Dirac, he did feel some unease. Not only did he write that justifying paragraph in the 1957 article but he also published a separate article[64] in *Nature* in 1961, in which he again tried to vindicate his result (based on a mistake!). During this process, he developed a rather unconvincing argument that was later the foundation stone of the so-called Brans-Dicke theory. In particular, he tried to argue against Dirac's second hypothesis concerning the number of particles, which Dicke, due to his mistake, had failed to reproduce. The title of his article, "Dirac's cosmology and Mach's principle" presumably annoyed Dirac, and he promptly defended himself by reaffirming the significance of the Large Number Hypothesis. He was, of course, right.

It is tragic that these two brilliant physicists, who had it in their hands to create a new cosmology by uniting their ideas, got locked in a petty dispute. The two articles certainly did not promote their cooperation and it is unknown if they ever talked to or met one another. One could even question if Dicke had read carefully the last part of Dirac's article of 1938, which similarly introduced variable scales. However, Dirac certainly did not scrutinize Dicke's paper but confined himself to the

Chapter 10: Einstein's desired cosmos

defense of his territory. Otherwise he would have noticed the close similarity of the approaches. Above all, he would have stumbled on the variable speed of light idea – the piece of the puzzle missing from his theory of 1938.

It makes sense to summarize the results obtained so far for a comprehensive view of the new picture of the evolution of the universe which could have arisen as a result of the ideas of Mach, Einstein, Dicke and Dirac.

Quantity		Unit		Epoch	Example	
Absolute time	t	s	t	10^{52}	10.000	↑↑↑↑
Speed of light	c	m/s	$t^{-1/2}$	10^{-26}	1/100	↓↓
Wavelength	λ	m	$t^{-1/4}$	10^{-13}	1/10	↓
Frequency	f	1/s	$t^{-1/4}$	10^{-13}	1/10	↓
Current time step	τ	s	$t^{1/4}$	10^{13}	10	↑↑
Speed	v	m/s	$t^{-1/2}$	10^{-26}	1/100	↓↓
Acceleration	a	m/s²	$t^{-3/4}$	10^{-39}	1/1000	↓↓↓
Inertial mass	m	kg	$t^{3/4}$	10^{39}	1000	↑↑↑
Cosmic horizon	R	s	$t^{1/2}$	10^{26}	100	↑↑
Observed horizon	R'	s	$t^{3/4}$	10^{39}	1000	↑↑↑
Perceived time	t'	s	$t^{3/4}$	10^{39}	1000	↑↑↑
Volume universe	V	m³	$t^{3/2}$	10^{78}	1.000.000	6↑
Number of particles	N	–	$t^{3/2} = t^2$	10^{78}	1.000.000	6↑
Mass of the universe	M	kg	$t^{9/4} = t^3$	10^{117}	10^9	9↑

Summary of the variabilities of physical quantities in Dicke's model. For the sake of simplicity, the exponents are rounded to integers; in reality t would be closer to the exponent 53 instead of 52.

Neither is the model in line with the Big Bang model in the usual sense nor does this picture deny an evolution of the cosmos, as Big Bang critics usually do. It is a synthesis in the best sense, not in the form of a feeble compromise, but in a surprising cross-connection that corresponds exactly to the observations. But how does this model alter the picture of the beginning of the universe, traditionally called the Big Bang?

ATOMIC NUCLEI DO MEAN SOMETHING

As mentioned, the cosmos does not really expand. Essentially, all matter is at rest, and only light, which can do nothing else but spread out, constantly expands the horizon of the universe that has grown to an incredible size. If one pursues Dicke's model to its logical end, then the size of the universe was, at some stage in the distant past, as large as that of the proton. One could label this state $t=1$ as a One-particle Big Bang, for the universe consisted at that time of just one elementary particle.[i] This means that the density of the universe at that time corresponded to the density of the atomic nucleus.

This is a dramatic difference from the conventional Big Bang when allegedly *all* of today's visible particles (10^{78}!) were concentrated in the volume of an atomic nucleus or within an even smaller space. This claim is already unscientific because densities exceeding that of the nucleus have never been reported by serious laboratory physics, despite the CERN folks who purport from time to time to have realized a "quark-gluon-plasma" or even (hilariously) a Big Bang simulation. Never was a higher density than that of nuclear matter observed over a significant time span or space volume. By the way, this is also true for astronomical objects – interestingly, pulsars have about the

---

[i] We can establish a semantic association with the idea of the "primordial atom" that was promoted by the Belgian priest Georges Lemaître. Lemaître played a significant role in the development of the prevailing cosmological model. The parallel is however artificial; Dicke's concept is radically different.

Chapter 10: Einstein's desired cosmos

same density as atomic nuclei, and whether black holes[i] with higher density exist is at least debatable.[ii]

In Dicke's cosmology, the value of nuclear density (which so far seemed to be an arbitrary quirk of nature), acquired a meaning for the first time. It established the deep connection between cosmology and atomic physics that Dirac had spoken of.

WHAT WOULD EINSTEIN HAVE SAID?

> "Your calculations are right but your physics dreadful" (Einstein to Georges Lemaitre, who advocated the Big Bang model)

In view of the fact that the cosmology presented here owes a considerable amount to Einstein's 1911 idea (if he had had the chance to compare it with cosmological data), we should take the perspective of natural philosophy and recall Einstein's thoughts about a cosmos governed by simple laws.

A static universe that existed for all time was undoubtedly Einstein's preferred model. When, after the discovery of the Hubble redshift, a static universe became unlikely, the idea of a steady-state universe gained ground: a dynamics without change of state, comparable to a uniformly streaming river. New research appears to show that Einstein even constructed the first steady-state model but then cast it aside.[65] The steady-state model was popular up to the 1960s. It allowed for a dynamics, but had to postulate strange mechanisms such as matter creation in order to avoid the Big Bang hypothesis. Despite the great difficulties of the modern Big Bang model ('concordance mod-

---

[i] The idea that a celestial body can become so massive that no light can escape from it became popular in the 1950s and was dubbed a "black hole." The idea however can be traced back to John Michell (1784).

[ii] It is a very unsettling and unintelligible feature of conventional gravity theories that very massive black holes with a very low density are postulated to exist.

el'), these alternatives were too exotic to find general acceptance. In addition, matter-creation models are not exactly the paragon of simplicity that Einstein desired. It was the classical situation in which the scientific community thought it had to decide between two competing models, with seemingly no alternative. Development of Einstein's idea of 1911 suggests that both the conventional model of the expanding cosmos and the steady-state theory were wrong, however.

Obviously, Einstein was intuitively reluctant to embrace the Big Bang model, but his distaste for an expanding universe was somewhat justified. In fact, a deeper reason *why* the universe should materially expand is lacking. Since Hubble's discovery was interpreted as an expansion, the world has got used to the arbitrary feature that nature had seemingly invented. In a more general sense, however, the Hubble expansion is an excuse, a complication in an epistemological sense: such ad hoc hypotheses are postulated to describe unexpected observations, while a fundamental reason for the effect's existence is missing.

Einstein's basic idea and Dicke's realization of it result in an amazing synthesis of the useful parts of the Big Bang and steady-state models: the universe is static, as far as matter is concerned, but on the other hand, it has a beginning which can be called the Big Bang ("Big Spread" may be better, perhaps) that simultaneously avoids many serious difficulties of the conventional model.

Equally mysterious is the fact that the universe at the time of the one-particle Big Bang consisted of nothing else but elementary particles – they filled the cosmos without spaces. This model is not only a proper description of the observations but is also remarkably simple. One can be sure that not only Dirac and Dicke, to whom it goes back, would be fascinated, but Einstein as well, who had a passionate interest in the cosmos, although by an unfortunate twist of fate it was not granted to him in time to comprehend the universe in its true size.

Chapter 11

Back Before Newton?

Why we need to question the notions of space and time

I have struggled with whether I should expect you to put up with this chapter. You might feel that it is too much of a musing on my part—an undue interruption of the story. How the ideas of Mach, Einstein, Dirac, Schrödinger, and Dicke can be combined in such a surprising way has been the essence of what I intended to communicate, and you might be curious to know what evidence for the idea can be found in the cosmological observations that I will address in Chapters 12 and 13.

I feel rather satisfied to have tracked back the ideas of these old thinkers, but I still have nagging doubts about how this picture can be extended. I do not think I have to withdraw anything, but the thoughts developed thus far seem to lead to a point where undreamt conceptual difficulties arise. It is as if we have just completed an exhausting mountain hike and then a huge cliff appears before us. I can only hope not to curb your interest in cosmology by adding these pensive reflections, but I cannot hide what bothers me during my continuous search.

This does, however, make it easier to argue that Einstein was engaged with constants of nature, and we shall get back to his considerations on physical constants. His 1911 idea about variable speed of light could have revolutionary consequences—first of all, it may explain the gravitational constant G. But does that put an end to the endeavor of understanding the laws of nature? Which questions will remain unanswered? If Einstein was right that no arbitrary numbers should occur in nature, we must ask about the origin of the remaining constants of nature. This problem ultimately touches on the concepts of space and time, a concern throughout Einstein's life:

> "The normal adult never bothers his head about space-time problems. Everything there is to be thought about it, in his opinion, has already been done in early childhood. I, on the contrary, developed so slowly that I only began to wonder about space and time when I was already grown up. In consequence I delved deeper into the problem than an ordinary child would have done."

Einstein's dictum "arbitrary constants should not exist" represents a philosophical conviction we can either agree or disagree with. As a matter of fact, the history of science has documented that it was precisely this attitude that led us to findings of which Homo sapiens can be proud. The ultimate consequence can only be to try to remove any arbitrariness in the laws of nature, a subject upon which Einstein has recurrently commented. With the phrase "I wanted to know if the Creator had a choice," Einstein insinuated that he did not believe in this choice—wanted to understand nature. Completely.

GETTING RID OF THE GRAVITATIONAL CONSTANT G

The geocentric worldview held by medieval astronomers contained so many arbitrary numbers that it was not exactly a parsimonious description of reality. Newton's law of gravitation simplified in a revolutionary manner, instead of dozens of parameters, that only one gravitational constant G reigned, which was evermore believed to be uniform all over the universe. In the past chapters, I tried to illustrate that an even greater revolution could have resulted from Einstein's flash of genius in 1911:

Chapter 11: Back before Newton?

to render the gravitational constant redundant, that is, to calculate it from the data of the universe.

This comes as a surprise insofar as you might think that the existence of G was essential for our system of units based on meter, second, and kilogram. In Chapter 2, I described the Planck units: $\sqrt{\frac{Gh}{c^3}}$ for meters, $\sqrt{\frac{Gh}{c^5}}$ for seconds, and $\sqrt{\frac{hc}{G}}$ for kilograms. Since the numerical values are unmeasurably tiny, these Planck units have led to a series of untestable esoteric concepts such as 'strings' and 'cosmic inflation.' It is indisputable that three constants of nature are necessary if one wants to obtain three units (i.e., meter, second, and kilogram). On the other hand, making G superfluous would imply that the unit kilogram is obsolete.

Indeed, inertia, being since Newton the intrinsic characteristic of all masses, allows us to quantify masses as inverse accelerations. Newton's second law, $F = m \cdot a$, is thereby incorporated as a definition in the system of the laws of nature. This is a situation that happened often in the history of science. Initially, more concepts than needed were established; then, a connection between these concepts was discovered ("law of nature"). Eventually, this law of nature turned into a 'definition' that allowed for a more economical system of concepts.

In earlier times, the notions of temperature and kinetic energy were completely distinct issues. Later, it turned out that temperature was nothing else than the average kinetic energy of a particle, and the "law of nature," $\frac{1}{2} mv^2 = kT$, was established. However, this relationship can be viewed as a definition of temperature, causing it to lose its independence and become a derived quantity. When applying an analogous argument to G, all the laws of nature can also be formulated without using the unit kilogram; instead, masses can be measured by means of inverse accelerations with the unit s^2/m.

Einstein's Lost Key

TWO POSSIBILITIES FOR REDEFINING THE UNIT OF MASS

While it is not clear how a fundamental theory should assign a distinct inverse acceleration to the unit of mass, one of the following numerical references to a "fundamental" acceleration of nature will inevitably arise. Basically, there are only two options. One takes the cosmological point of view and obtains an acceleration by means of the term c^2/R_u, R_u, denoting the radius of the universe as usual (i.e., plain words, an acceleration that had speeded up objects to the velocity of light since the Big Bang. Alternatively, c^2/R_u can be viewed as the centripetal acceleration necessary to keep the most distant masses in the universe on a circular orbit with the velocity c.[i] When considering $c^2/R_u = 7 \cdot 10^{-10}$ m/s^2, it would be obvious to link the (huge) mass of the universe to this very low acceleration. It is appealing that this acceleration appears just where the observational anomalies related to the phenomenon of 'Dark Matter' show up, particularly at the edges of galaxies ("rotation curves").

However, I believe that it is more fundamental to concentrate on the microcosmos when redefining the notion of mass. In Chapter 9, we have already encountered the puzzling coincidence $h \approx c \cdot r_p \cdot m_p$ and recognized it as a form of Dirac's Large Number Hypothesis. Imagine a proton (or neutron) as an object with a radius r_p that rotates at its outer edge with the speed of light c. Here again, there is an orbital acceleration $c^2/r_p = 10^{32}$ m/s^2 that you would naturally relate to the proton mass $1{,}67 \cdot 10^{-27}$ kg.

There is even another justification for the latter approach. The spatial change of the speed of light (more precisely, the gradient of ¼ c^2) determines the local *gravitational* acceleration, but if

---

[i] Of course, this is an entirely hypothetical and not necessarily reasonable scenario.

Chapter 11: Back before Newton?

we take Dirac's reference to elementary particles seriously, it could also be that this acceleration is a manifestation of other forces, in this case, the nuclear one. Here, the nuclear force would turn out to be an unrecognized form of gravitation.[i]

There is one final intriguing aspect of this simplification of units. If one replaces kg with s^2/m, then the unit of Planck's constant h (which is $kg\ m^2/s$) accordingly becomes $m \cdot s$, the product of meters and seconds. Thus, the unit of length (meter) could be determined via $\sqrt{hc}$ and the unit of time (second) by means of $\sqrt{\frac{h}{c}}$. Where previously three constants gave birth to three units (meters, seconds, and kilograms), this number is now reduced to two.

The quantum of action h is very well known due to its prominent role in Heisenberg's uncertainty relation, which forbids the position and momentum of a particle to be measured at the same time. The uncertainty relation normally links terms such as position/momentum or energy/time, that is, any two quantities of which the product has the same units as h. The new unit system—assigning $m \cdot s$ to h—suggests yet another simplification: Heisenberg's famous law would then turn into a space–time uncertainty.

NEWTON DID NOT PREDICT C

Once we realize that the unit kilogram can become superfluous, the question arises as to whether the remaining units meter and second have the right to exist.[ii] But can we really hope to

---

[i] Before nuclear physicists scream with horror: Heisenberg, for instance, was convinced that something was wrong with the traditional understanding of nuclear force (Ann. Ph. 32; 1938, p. 29).

[ii] Ilse Rosenthal-Schneider, in her correspondence with Max Planck, also expressed the visionary idea to make additional constants of nature redundant (p. 41).

replace these everyday perceptions of reality by something "better?" Apparently, meters and seconds are deeply interwoven with the existence of the constants of nature c and h. While three constants of nature, G, h, and c were responsible for the system of units comprised of meters, seconds, and kilograms, meter and second still require two: c and h.

Thinking about this possibility is irritating. Can we ever practice physics, ever hope to make measurements, without expressing results in units such as meters and seconds? Meter and second are the innate attributes of space and time, the basis of our physical understanding since Newton. On the other hand, they are inextricably linked to the existence of h and c. Meter and second therefore are necessary units for Newtonian physics, but are the existence of h and c also necessary for Newtonian physics? Certainly not! Newton knew nothing about h and little about c, and he did not attribute something fundamental to the existence of the latter. Newton's physical legacy, the grand logical building of laws of nature, exists without these constants of nature—h and c are alien elements to Newtonian physics.

Even if it might be hard to admit, we have to take note of the incompatibility of Newtonian physics with these two constants. Newton defined the notion of velocity precisely as the temporal change of position, but that nature had singled out a distinct value for it was surprising. In particular, the finiteness of c had to appear odd from that perspective. In this respect, Ole Rømer's famous observation of the motion of Jupiter's moons in 1676, from which he calculated the numerical value of c with astonishing accuracy, constituted a conceptual problem, yet a finite speed of light could still be understood as an astronomical curiosity.

A more serious attack on Newton was Einstein's special theory of relativity, in which he also identified c as maximum velocity that no material object could exceed. This affected a pivotal part of Newtonian findings, namely, how quantities such as

Chapter 11: Back before Newton?

time, position, velocity, acceleration, mass, and force were related to one another. There is no reason at all in Newtonian dynamics for c to represent a limiting velocity. With a sufficiently high kinetic energy, a particle should be able to become infinitely fast. Experiments showed that this was not the case, and Newton's equations become less and less true when approaching the speed of light. This is a classic case of an anomaly, and also a bad sign, when we take a long-term, epistemological perspective, as American philosopher of science Thomas Kuhn did in his book *The Structure of Scientific Revolutions*. Anomalies are the subtle symptoms of a diseased scientific theory. With the discovery of the speed of light, Newtonian physics was no longer completely healthy.

ANOTHER BLOW FOR NEWTON

Differently, but in Kuhn's sense, even more clearly, Planck's constant presents an anomaly. Since you might still have difficulties imagining what this puzzling constant of nature actually means, I shall explain it to you in a little more detail.

Its physical unit is that of angular momentum ($kg\ m^2/s$ is the product of mass, velocity, and distance from the rotation axis). An easy way to grasp this is as follows: electrons that orbit the atomic nucleus always bear multiples of h as angular momentum. This great discovery by Niels Bohr was preceded by Einstein's formula $E = h \cdot f$, showing that h meant something profound not only for matter but also for light. However, a deeper reason for the existence of h physics is unable to give to this day, though it has been shaken to its very foundations by h.

The experimental results of black body radiation, ingeniously summarized from two other laws by Max Planck in 1900, obviously presented a contradiction to conventional physics. Later, it became clear that h did not allow for arbitrarily small lengths and time spans. Werner Heisenberg succinctly formulated this fact in his uncertainty principle ($dx\ dp = h$ and $dE\ dt = h$), pointing out that 1) infinitesimally short time scales lead to infinitely high energies, and 2) infinitesimally small lengths lead to infinitely high momentum. Obviously, neither consequence makes any sense.

Niels Bohr (1885-1962) and Max Planck (1858-1947)

This failure of Newtonian physics affected its very own domain: differential calculus. Newton,[i] before formulating the laws of dynamics, first had to develop a suitable branch of mathematics (a merit that has gone unmentioned thus far) in order to define terms such as velocity. This branch of mathematics is called infinitesimal calculus because it concerns itself with the infinitely small. However, this is exactly where the laws of physics as developed by Newton fail to deliver—in the case of small lengths and short periods of time. h must therefore, from an epistemological perspective, be seen as an anomaly that indicates that the terms of space and time are ultimately an inadequate description of reality. It is astonishing, and prob-

---

[i] Leibnitz also developed a form of infinitesimal calculus.

Chapter 11: Back before Newton?

ably to be understood only by the long-time scales of scientific evolution, that this aspect was hardly ever emphasized.

Mathematicians in particular never seemed perturbed by the fact that elementary notions such as continuity and differentiability had suddenly become irrelevant. The current, amazingly superficial paradigm of physics is that in the case of time and length scales, "quantum-mechanical corrections" are necessary, or "Heisenberg's uncertainty relationship is to be observed" – what if you please does this explain? Leibnitz' dictum[i] "*natura non fecit saltus*" is simply the statement that nature should obey mathematics. Many mathematical theorems used in physics require the concepts of continuity and differentiability if they are to be proved, yet quantum mechanics did not care to overturn these concepts without being able to specify a reason for the existence of h, which was grounded on mathematics.

If we are honest and take Einstein's belief in the simplicity of the laws of nature seriously, the appearance of the phenomena h and c can only be seen as a falsification of Newtonian physics. In the generation-long periods of time in which scientific knowledge has evolved, this was concealed by findings that, in themselves, were certainly brilliant.

> *The normal mathematician, even if he is good, understands nothing of physics.* –Werner Heisenberg

> *If we have to go on with these damned quantum jumps, I regret that I ever got involved with quantum theory.* –Erwin Schrödinger

---

[i] "Nature does not make jumps," meaning that all dynamic processes ought to run continuously.

Einstein's Lost Key

THEORY OF RELATIVITY: THE LARGE-SCALE REPAIR

Around 1900, when it became clear that the speed of light also played an important role in the dynamics of material bodies, physics faced a confusing situation. Lorentz and Poincaré provided decisive contributions to relativity, but Einstein cut the knot as he elegantly established new laws that not only described the observations perfectly but also preserved Newtonian physics as a valid approximation for small velocities. The enormous value of special relativity lies in its conceptual simplicity; it describes a series of experiments, such as time dilation, length contraction, mass increase, and many others, with only one constant: c. In this way, the speed of light acquired a fundamental importance for both light and matter.

In view of the current standard models of physics that deal with a confusing number of arbitrary parameters, it may seem presumptuous to be satisfied with the simplicity provided by just two constants of nature. However, c is, as viewed from Newtonian physics, an unexplained parameter, and it could be that it is one too many. Just introducing new parameters whitewashes the incomprehension of a phenomenon. Today, even the use of c in the laws of nature could be a subtle dressing up of misunderstanding. As brilliant as his theory of relativity was, Einstein repaired Newtonian physics in order not to shake its fundamentals. In the end, he protected the Newtonian concepts of space and time. In Kuhn's sense—alas, the history of science applies to such long periods—it must be said that Einstein only refined the paradigm of space and time that Newton had developed. Perhaps he should have questioned it in its entirety.

Certainly, the ongoing mental effort in the study of relativity won't be lost, already today it is fruitful and of lasting value for mathematics. However, will it assert itself in the worldview of physics of a distant time, which has to fit into a

Chapter 11: Back before Newton?

world broadened by manifold new insights; will it have any significance in the history of science beyond being a spirited Aperçu? –*Ernst Mach*

QUANTUM THEORY: THE SMALL-SCALE REPAIR

Equally important as Einstein's theory of relativity is the theory that integrated Planck's constant h into physics: quantum mechanics. Although in this case, a key contribution came from Einstein, quantum theory is unthinkable without the work of Planck, Bohr, Heisenberg, Schrödinger, and Dirac. The manner in which atomic physics was revolutionized at the beginning of the 20th century by the appearance of h is unique. For classification purposes, it also helps to look at Kuhn's description in terms of paradigms. The process of simplifying the laws of nature by fewer parameters became repeatedly visible. Planck himself constructed a straightforward synthesis of the radiation laws of Wien and Rayhleigh-Jeans, and the triumph of his constant h continued by generating other revolutionary simplifications. Einstein realized that h was not just a curiosity of black body radiation, but that it was a fundamental property of light: energy could only be released in certain portions ('quanta'), or $E = hf$ – by the way, an interpretation heavily opposed by Planck. In this respect, Einstein contributed more to the recognition of h than Planck himself.

Finally, Niels Bohr contemplated the units of the constant h, and in a stroke of genius, recognized its importance for atomic physics. Electrons orbiting the atomic nucleus always had to have multiples of h as angular momentum! This was the climax of quantum mechanics (Heisenberg and Schrödinger developed mathematical formalisms that justified Bohr's approach), and more simplifications that eliminated further constants of nature were to follow. Johann Jakob Balmer's number,[i] discovered in 1885 in the spectra of the hydrogen atom, was explained by a formula that contained h.

REVOLUTIONARY DISCOVERY OR ADHERENCE TO CONVENTIONS?

These findings were sensational; they were a wonderful description of the phenomena caused by the action quantum h, not to mention the technological impact they had on our entire civilization. Yet the impressive building of quantum mechanics ultimately obscured the fundamental problem: the incompatibility of its concepts of space and time with Newtonian physics. In fact, quantum phenomenology falsified Newtonian physics.

Just as in the case of c, Newton certainly did not predict the appearance of h. In the same manner that the speed of light presents an upper limit beyond which Newton is not applicable, h is a lower limit, below which the conventional laws of motion lose their validity. In Newtonian mechanics, there is no reason preventing lengths, times, speeds, energies, etc. from becoming arbitrarily small.

It may sound overbearing to call the possibly greatest achievements of physics, relativity, and quantum mechanics unsuitable constructs. And it is wrong in the sense that the merit of these theories consists precisely in reducing a variety of natural phenomena to a few simple concepts. Two constants, h and c, in this sense, are a very modest toolbox. But, if we appreciate the progress up to this point, we cannot overlook the natural consequence: If c and h prove themselves obsolete, they would prompt an even deeper understanding. Further, this would lead

---

[i] Now called the Rydberg constant, it has a value of $1.09737316 \cdot 10^7$ per meter.

Chapter 11: Back before Newton?

to new concepts that are more suitable for a fundamental description of nature than space and time.

Howsoever, one cannot deny that h and c represent a complication of the laws of nature. Without quantum mechanics and relativity, physics was simpler—albeit wrong. A commonplace in today's point of view is that a unified theory of physics requires relativity and quantum mechanics to be linked, even though both theories are conceptually and formally alien to each other. However, the problem may be ill posed: Presumably, it is not a unification that is needed but a reconstruction that replaces the concepts of space and time by more adequate ones.

Seen from a long-term perspective, we find ourselves in a bizarre situation. Physics has developed a frame of concepts (time, length, velocity, acceleration, force, mass), and with its powerful logical structures, nature's surprising secrets have eventually unveiled inter alia constants of nature such as h, c, and G. During this process, an obvious connection surfaced between the number of physical dimensions and the number of constants of nature. Advances in understanding led to a reduction in the number of fundamental constants (although contemporary physics moves just in the opposite direction), therefore offering a more economical system of concepts (remember that energy substituted temperature). But what is the end game? Is it possible to develop an even better understanding of physics and manage with even less constants? Certainly, we cannot get to this point without touching on the concepts of space and time. Apparently, each such "physical" notion is already a piece of incomprehension. We shall understand the laws of nature only when the last of these quantities involved with physical units has become obsolete. True understanding will be obtained only when the basic notions are justified directly from mathematics.

In view of the critique of futile mathematical constructs in contemporary physics, I have uttered elsewhere such a statement may seem paradoxical to some. However, I do believe that the true mathematical connections that nature employs are not yet discovered.

GETTING ALONG WITHOUT SPACE AND TIME?

Exasperating consequences await us if we try to think about replacing the notions of space and time. These two very basic concepts are so deeply rooted in our daily experience that we cannot imagine, even if we want to, getting rid of them. Above all, however, such a perspective shakes the axioms of physics itself. Not just current and modern physics (i.e., circa 1900) would be at stake, but the whole of science up to now would be called into question. Our entire system of physical quantities goes back to Newton, and perhaps it was disastrous that he did not delve into two of them: space and time.

The connection between the two remaining units meters and seconds on the one hand and the constants h and c on the other hand is obvious. Meters result from the combination $\sqrt{hc}$, while seconds result from $\sqrt{\frac{h}{c}}$. If, however, h and c have origins that are not understood, then we definitely know that we have not comprehended space and time. Under no circumstances may we take these two concepts for granted just because they are so easy to experience in everyday life; by using this argument, one could also justify all primitive descriptions of natural phenomena. And of course, this consideration is not to criticize Newton—as indeed he created the conceptual frame from which physical science was born. However, Newton could not foresee the existence of the velocity of light c and of the quantum of action h. All the more, we must take seriously the fact that his observations forced us to take note of these two constants, contrary to the expectation at the time—Newton's laws described physics perfectly. Both h and c are therefore anomalies in the

Chapter 11: Back before Newton?

sense of Thomas Kuhn's theory. The paradigm at risk is as elementary as never before: space and time.

AN UNKNOWN WORLD OF NEW CONCEPTS

> *The book of nature is written in the language of mathematics.* –Galileo Galilei

This is where one might to start from: In which field of mathematics are there algebraic structures that show a behavior characteristic of h? Indeed, there are such structures—for instance, the nonlinear effects when concatenating rotations in three-dimensional space. But mathematics has not yet revealed a structure that by sheer logic delivered an object with the properties of h. Einstein's reiterated postulate, "I wanted to know if the creator had a choice," has remained unfulfilled up until now.

Einstein's critique of quantum mechanics, alas, led instead to an underestimation of the significant role he played in the discovery of h. By the way, his famous words, "God does not throw dice," were not generally directed against randomness. He simply insisted that a justification had to be given from first principles as to why calculations involving elementary particles should not be deterministic: "I'm not saying, *probalilitatem esse delendam, but esse deducendam*."[i] It was Einstein who, against the express opposition of the discoverer Planck, bestowed the quantum effect with its fundamental role in physics. The two big puzzles that physics still has to solve, h and c, are therefore connected with Einstein's name. However, it is likely that New-

---

[i] "I don't say probability is to be destroyed, it is to be deduced however." This was an allusion to the saying of Cato the Elder "*ceterum censeo Carthaginem esse delendam*" in ancient Rome; he continuously demanded the destruction of Carthage.

tonian physics of space and time have to be overturned even more thoroughly than was possible for Einstein to do.

The mystery of quantum theory is often illustrated with the "dualism" of waves and particles, a concept developed by Niels Bohr. In fact, it is amazing that some experiments, such as electron diffraction in crystals, clearly speak for the wave nature of matter, and in certain others, such as the photoelectric effect (Einstein won the Nobel Prize for it), light behaves like a particle. A flood of publications have emerged that debate the various interpretations of quantum mechanics, focusing on this wave–particle dualism. But waves and particles are probably simply ill-defined terms that touch on our traditional idea of space and time.

Truly wondrous, however, is the fact that the world presents itself in such contrasting phenomena as light and matter. In my opinion, this is the real puzzle that awaits an explanation. Needless to say, it is clearly connected with the riddle of the existence of fundamental constants: c belongs to light, just like h belongs to matter. Today, we still do not have an explanation for either pair.

Chapter 12

Forgotten But Not Hidden

Einstein's idea is visible long since: It is dark energy

This chapter shows that crucial cosmological observations support Einstein's idea of variable speed of light. It also illuminates how difficult it is to correct existing misconceptions that have been firmly anchored in the scientific community for decades.

The most important cosmological discovery in recent times is the so-called 'accelerated expansion of the universe,' also well-known as 'Dark Energy,' a hypothetical substance postulated to explain the acceleration that up until then, nobody could account for. As questionable and as suspicious as these terms might be, a significant discovery lay behind them.

In the 1990s, two groups of researchers became particularly interested in supernovae. These stellar explosions are extremely rare—we can expect one about every one hundred years in the Milky Way. Due to the sheer numbers of galaxies observable by devices such as the Hubble space telescope, we can now hope to find hundreds of supernova explosions each year. Scientists developed a tricky method for a systematic detection and were soon able to discover an impressive number of supernovae in the vastness of the universe.

These explosions are of fundamental importance for cosmology because they allow scientists to take relatively precise distance measurements. Astronomers are fairly confident about their understanding of the conditions under which stars explode, and they therefore can give estimates of the absolute brightness of such an event. For example, if there is just a weak glow visible through the telescope, the exploding star and the respective galaxy must be very distant. Nevertheless, the explosions are not all equally bright, spoiling the precision of the distance measurements.

However, one researcher in the two rival groups had the brilliant idea to compare the temporal evolution of the brightness of the different explosions ('light curves'). It turned out that the brighter explosions lasted longer, whereas supernovae with weaker luminosity often waned after a few days.[i] Independently of a thorough understanding of the explosion process (which is still missing), they thus managed to achieve a new level of precision in cosmological distance measurement. Deservedly, the two groups were awarded the Nobel Prize in 2011. The results became well known because they were a complete surprise: the data grossly deviated from the expectations of the standard cosmological model.

COSMIC BEACONS ARE NOT AS BRIGHT AS EXPECTED

What was the cause of this contradiction? One has to realize that distant supernova explosions represent a look into the distant past, since their light takes billions of years to reach our telescopes. The observation of supernova explosions at varying distances bears witness to the dynamic development of the universe: every single explosion provides information about the expansion rate according to the model, and the sum of the observations delivers us a sort of film in which we get an overview of the entire evolution of the cosmos.

---

[i] The brightest supernova in the history of astronomy occurred in AD 1054 and was visible for a whole month during daylight.

Chapter12: Einstein's idea is visible long since

Obviously, this film did not meet the researchers' expectations, as distant supernovae appeared too faint. However, if we assume that from a certain point in time, the universe developed a repelling force that drove its galaxies further apart, we are again in line with the data. The term 'accelerated expansion' evolved out of this. Why was this conclusion premature?

If a result does not match the expectations, it is an anomaly, often about a new effect. But it is also possible that the expectations were based on incorrect assumptions. One rock-solid conviction of cosmologists was of course that the universe expands. Eighty years after Edwin Hubble's discovery, it was literally impossible to scrutinize such an assumption.

However, in this conventional picture, the expansion should be held back by the gravitational pull of the masses, notwithstanding their expansion; this had been accounted for in the calculations upon which the predictions were made. But according to the Einstein–Dicke model, the cosmos does not expand at all, since the redshift was caused by a decreasing speed of light that reflected the aging of the universe. Consequently, there is no reason for why this illusionary expansion should be restrained. The gravitational force does act in this model, but not in an unnatural global manner—generating local mass concentrations was the only thing it could do.

The alleged acceleration of the expansion is therefore nothing else than the absence of deceleration. The hypothetical brake is not there, a downright simple explanation! Did the supernova researchers consider such a simple mechanism? Yes and no. To interpret the alleged acceleration of expansion as the absence of slowdown was so obvious from the data that it even found its way into some publications as the 'empty universe model.' However, such a model would seem absurd in the conventional perspective because it lacked any theoretical justification. We could argue that the universe is anything but empty; thus, it makes no sense that the expansion appeared as if there were no masses. Although data obviously supported the 'empty universe model,' it was never seriously deliberated. The dark energy interpretation won the day. The notion of a universe without galaxies, stars, gas clouds, simply nothing at all, was too crazy to be possible.

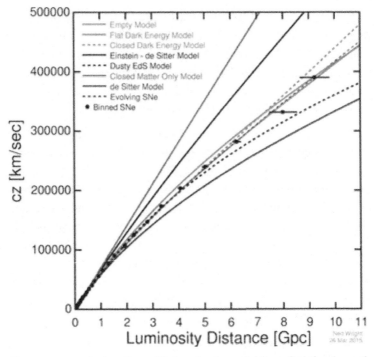

Supernova explosion data. Obviously, the solid line slightly above the points (empty model) is also compatible with the observational data, and it is of brilliant simplicity. [66] *But because there is no explanation for an "empty universe" in conventional thought, the hint is not taken seriously. [Ned Wright]*

The very simple model based on Dicke's thoughts clearly corresponds better to the phenomenon than the complicated combination of a real expansion—the presumption of deceleration by conventional gravity combined with acceleration by dark energy. But since the latter picture 'accurately' reproduces the

Chapter12: Einstein's idea is visible long since

data, we are back to the idea that Einstein will have a hard time, to put it mildly, as a few thousand, if not tens of thousands of astrophysicists, having accepted dark energy (albeit initially reluctantly) is not easily challenged. Although recent data suggests that the evidence for the `accelerated expansion' is in any case much weaker than claimed has done little to diminish belief in the standard model.

LAMBDA: WATCH OUT!

If a key message of this book is that the mystery of dark energy was solved by one of Einstein's age-old ideas, this could lead to a grandiose misunderstanding, against which I have to issue an explicit warning. Einstein had indeed unfortunately forgotten about his own idea of 1911, or rather, he had overlooked its cosmological relevance. But that did not mean he was not also dedicating fundamental thought to the cosmos. It must have been an unpleasant surprise for him that the equations of the geometrical version of general relativity, applied to the cosmos as a whole, did not result in such a simple solution as would have suited his philosophical convictions about nature.

Einstein had a particular predilection for a static universe, existing forever in the same shape. This possibility, however, was ruled out by his own equations, as the Russian mathematician Alexander Friedmann demonstrated in 1917. In the same year (long before the beginning of observational cosmology), Einstein added a so-called cosmological constant Λ ("lambda") to his equations. Without meaning disrespect, I would say that this was a patch in order to save his idea of the static universe. Later, he blamed himself ("my greatest blunder"), and today's historians are eager to remark that he should have predicted the expansion of the universe instead, which Hubble discovered in 1930. But perhaps Einstein's reluctance against the expansion of the universe was well justified. It is the inherent arbitrariness in modern cosmology that has thoroughly suppressed any questioning of the unknown origin of the expansion.

In any case, the cosmological constant Λ fell into oblivion before it was brought up[67] again in 1998—ironically, as a justification for 'dark energy.' Some prominent cosmologists saw a vague mathematical parallel (though the physical units do not match) to Λ, and promptly, an article was published in an issue of the journal *Nature*, which featured a pipe-smoking Einstein on the cover. It is not hard to understand that this analogy was diametrically opposed to his philosophical convictions: Einstein put in a mathematical complication that showed his desire for the physical simplicity of the laws of nature; he just could not see how to do it otherwise. In contrast, Λ, or 'dark energy,' is also a conceptual complication that would certainly have been repugnant to him. Yet Einstein has since become an (abused) key witness to the idea. The key to an explanation of 'dark energy' is not to be found in his idea from 1917 but in a completely different idea that goes back to 1911: variable speed of light.

The modern interpretation of the cosmological data is a prime example of how opinions are formed in the research community: their beliefs were so deeply engrained, they could not see the forest for the trees. In Hubble's days, when the redshift of galaxies came as a surprise, its interpretation as a Doppler shift of objects that moved away from us seemed to stand to reason. The term "receding velocities" remained dominant in the literature on the subject for about thirty years. In the meantime, the question that ought to be asked as to the deeper reason for the expansion slipped into the background, a typical phenomenon when investigating the sociology of science.

GLOSSING OVER CONTRADICTIONS

Much later, it turned out that postulating an expanding motion was not sufficient to explain the observed data, which was, in fact, another alarm signal. In 1965, an unexplained residual

Chapter 12: Einstein's idea is visible long since

signal of a microwave antenna observing the sky was interpreted as a remnant from a hot, dense state of the early universe.[i] Meanwhile, three space missions, COBE, WMAP, and Planck, have taken pictures of the signal all over the sky with impressive precision. However, this so-called cosmic microwave background heavily contradicts a Doppler shift interpretation of emitting light sources. If they really moved away from us, the lights would be weaker by a factor of 1000. If scientists had properly followed the standards of scientific methodology, the model of an expanding universe would have been considered falsified.[ii]

But, as is so often the case in history, researchers were stunned at first and then made additional assumptions that enabled the model to digest the data. Today, it is presumed that it is not the objects in space that are moving away from one another (this would correspond to a velocity) but that 'space itself' is expanding. What does this mean? I refer here to an invisible construction: signals usually coming from matter only and not from 'space itself.' Referring to matter that expands, imagine a space in which this happens. But where does space itself expand? In another space perhaps? All of this seems to be jiggery-pokery, using definitions that help to hide one obvious contradiction: matter does not move apart (in the same way as parts of galaxies do that are known not to extend), but light is shifted to the red by a 'space expansion' while it is traveling. Ultimately, this means the invention of a third entity in addition to light and matter: the ominous 'space' attributed with independent proper-

---

[i] You might notice that I am mentioning this with some reservation. The criticism of the data evaluation by Pierre-Marie Robitaille certainly contains some valid points.

[ii] Here, I refer to the concept of falsifiability developed by Karl Popper. A further anomaly in this context is the so-called 'Tolman surface brightness of the galaxies.'

ties, a mathematically embellished complication of the model that has no scientific justification.[i]

Many people have gotten used to accepting exotic constructions for describing reality. They view physics as a pleasant stroll through a botanical garden and look forward to the phenomena without feeling bothered by the fact that they do not understand them. However, the complication that these new concepts bring with them is undeniable. Epistemologically, this is a symptom of sickness.

How much easier would the solution based on Einstein and Dicke be? There is only the space defined by matter (the idea of Ernst Mach); no expansion; and consequently, light alone is spreading and is thereby necessarily shifted to the red. Everything fits into a coherent, logical picture. However, one big obstacle remains: the scientific community would have to discard many of their long-established convictions.

A COINCIDENCE PROBLEM AND A TWO-HUNDRED-YEAR-OLD PUZZLE

A rather odd anomaly is known as a coincidence problem. Already postulating an acceleration that starts more or less at a certain moment of the cosmic evolution seems rather contrived. However, an additional strange consequence arises if one calculates the age of the universe backward based on the expansion rate.

The theoretically assumed deceleration and the alleged acceleration of the expansion compensate each other just so that the current expansion rate—seemingly at random—perfectly matches the age of the universe. Of course, this is just another

---

[i] Further serious problems of the notion of space expansion were discussed by Baryshev (arXiv.org/abs/0810.0153)—sadly, his articles had few followers.

Chapter12: Einstein's idea is visible long since

aspect of the aforementioned mystery of "flatness," but only the Einstein–Dicke cosmology allows us to make a connection between the 'coincidence problem' and the 'empty universe.'

In addition, there is a much older fundamental problem of cosmology, which also would have been solved if Einstein had returned to his idea of 1911. It is closely connected to the issue of kinetic and potential energy covered in Chapter 10. To this day, we do not understand why two such different forms of energy exist. If we contemplate the universe as a whole, a spectacular conspicuousness comes along that we have already encountered as "flatness": kinetic and potential energy are approximately of the same amount. For decades, this was the dominant topic in cosmology. Why?

All the masses in the universe contain, due to their mutual attraction, potential energy, and the question arises as to whether kinetic energy contained in the assumed expanding motion is larger or smaller than the potential energy. In the first case, the force of gravity that pulls together the universe could never completely slow down the movement of the diverging galaxies, and the cosmos would expand forever. This was called an 'open universe'; conversely, a universe is called 'closed' when its expansion eventually comes to a standstill, and as a result, collapses in itself. This would happen, so it was assumed, when the potential energy outweighed the kinetic energy. Finally, cosmologists describe the intermediate state, implying an equal amount of the energy forms, as "flat."

For a long time, observers disagreed on which version was now realized in the cosmos; this on its own is definitely noteworthy. How unusual the equality of the two forms of energy was only became clear thanks to the considerations of a deep thinker who you already know: Robert Dicke. He published his thoughts regarding the open and closed universe in 1969 (i.e., a decade after he had given up his visionary approach of 1957). Dicke calculated what the temporal evolution of an open and/or closed universe would pass off and came to a surprising result: if the kinetic energy was only a tiny fraction greater than the potential energy, then the universe would have expanded so much since the big bang (he assumed the conventional model) that the measurements would have clearly revealed an open universe. Had, conversely, potential energy prevailed by a minuscule amount, the expansion would have already come to a halt, and we would observe a contractive universe. With this, no one would have ever raised the question of "open" or "closed."

DICKE AGAIN

Today's measurements lie pretty close to the case of the "flat" universe. What does this mean for the early state of the universe? Dicke calculated that initially, potential and kinetic energy must have been balanced to sixty (!) decimal places. They coincided exactly and made for a 0.000000000000000000000000000000000000000000000000000000000001 percent match! As rational human beings, we can hardly believe in such a coincidence. The only reasonable conclusion is that our distinction between kinetic and potential energy in the cosmos must be based on a misunderstanding. For some unknown theoretical reason, both kinds must apparently have *exactly* the same amount. This is the result he would have gotten, had Dicke taken his own idea of 1957 seriously. In the variable speed of light model, Einstein's formula $E = mc^2$ (which is already famous enough) has an even deeper meaning: the potential energy is lowered in gravitational fields, and this is expressed by a lower (square of the) speed of light. When an object enters such a gravitational potential, it compensates for the loss of potential energy (a lower c^2) by a correspondingly higher mass m (according to special relativity), and the total energy remains the same. The distinction between kinetic and potential energy is thus obsolete, as their equality would by definition be incorporated into the formula $E = mc^2$. Combining Einstein's idea from 1911 and Dicke's addition from 1957 would thus

Chapter12: Einstein's idea is visible long since

have solved one of the greatest mysteries of cosmology, and the problem would never have been raised. Particularly incomprehensible is that Dicke, who was the first to point out this serious contradiction in 1969, did not remember his own brilliant idea.

Below, I shall briefly address another, somewhat unpleasant, aspect of the concept of the 'flatness' of the universe. By the early 1970s, cosmology was infiltrated by a number of particle physicists who were keen on extrapolating the laws of nature into unknown territory. Depending on your perspective, this can be called bravery or megalomania. The so-called inflation theory is based on the idea that the cosmos expanded over a period of 10^{-35} seconds right after the big bang at multiple superluminal velocity. While there is no hope of observing this exotic scenario anywhere or anyhow (a lack of testability that should lead to an immediate dismissal of a theory if physics was healthy), it is still promoted as 'cause' for the flatness of the universe. Alas, there is even a nonzero risk that, due to endless perpetuation and awards sitting on the shelves backstage, inflation will gradually be granted the rank of an "experimentally verifiable theory."

I do see a glimmer of hope in this bizarre situation, however. The more blatant the nonsense, the sooner an educated public will defy the expertocracy of the scientific community and question their dogmas. For instance, how much intellectual credibility would you concede to a foundation based at Yale University that awards a $500,000 prize to proponents of "multiverse" theories during "cosmic inflation?" I can only hope that up to this point, you have already developed a certain gut feeling that might help to distinguish the content of one of Einstein's sound ideas from modern bragging.

> *In order to be a distinct member of a flock of sheep, you have to be a sheep in the first place. –*
> *Albert Einstein*

THE BUSINESS OF BIG SCIENCE

*The voice of the majority is no proof of justice. –
Friedrich Schiller*

Despite these excesses, there are still a large number of honest and skilled scientists who are convinced of the correctness of the current model. I shall open a parenthesis here to reflect the situation that emerges for you readers—that is, when a popular book like this completely goes against the interpretation of cosmological data accepted by the majority of researchers. Are such entirely dissenting arguments necessarily outlandish?

Typically, a firmly established interpretation such as the expansion of the universe is formed over generations, and for this very fact, it is hardly subjected to scrutiny. In principle, this may suddenly change. The principle of majority may make sense in politics, but in scientific research designed to discover the truth, it is of little use. Research communities have often misinterpreted observations.

Whoever is aware of these historical facts also sees that the idea of an idealized science, in which observation is the judge that categorizes theories as 'right' or 'wrong,' is also highly naive. Even if there is an objective core in science, it is almost always an interpretation of data and the general agreement upon it that forms our picture of reality.

This is a central topic in Thomas Kuhn's book *The Structure of Scientific Revolutions*. Though nobody has ever given sound counterarguments to Kuhn, his theses are only superficially reported in research, and most practitioners ignore their central point. Kuhn has described so unequivocally the mechanism of how scientific paradigms become eroded by anomalies of observation, arbitrary assumptions, and gradual complication by means of free parameters, that every rational being would recognize these patterns in contemporary cosmology. The fact that

Chapter 12: Einstein's idea is visible long since

the researchers involved are usually blind to this kind of insight was something that Kuhn had foreseen.

A popular criterion by which the scientific mainstream attempts to distinguish itself from alternative opinions is via "publication in a recognized scientific journal." The failures of this peer review system have been so thoroughly discussed[68] that they need not be repeated here. Beyond this, the argument is simply wrong—there are many published results that criticize the current model, offering alternatives, but as long as the majority does not listen, the caravan rolls on.

For example, the famous astrophysicist Fritz Zwicky (1898–1974) had already noted that there was no evidence for an attraction between galaxies,[69] but since the claim did not fit into any schema, the article was never quoted. Zwicky's observation remains a curiosity that causes scientists to shrug their shoulders, but it does not push them to any fundamental reflections. It is the same with galaxies being too small, the coincidence problem, or the mysterious flatness of the universe. Ultimately, researchers do not know what to make of these phenomena in times of "normal science," as described by Thomas Kuhn. Enigmatic connections eluding any immediate and cheap explanation simply offer too little opportunity for satisfying the ever-present demand for publication.

> *If one had taken the ideas of these scientific geniuses who were the founding fathers of modern science and submitted them to committees of specialists, there is no doubt that the latter would have viewed them as extravagant and would have discarded them for the very reason of their originality and profundity.*[70] *–Louis Victor de Broglie, Nobel Laureate 1929*

SOCIOLOGY AND PARALLELS TO THE DARK AGES

Rather than pondering the problems behind, it is much easier to establish concepts such as 'dark energy' (however arbitrary they might be) and then keep looking for them elsewhere. The mutual recognition of the phenomenon brings more profit for all parties involved than the 'thwarting' of such 'progress' by cultivating epistemological doubts. Only by taking a historical perspective can we comprehend a phenomenon that in the meantime has seen entire research programs set up to examine dark energy in various places, and of course, eventually find it.

Thus, dark energy is now regarded as 'independently verified' by the distribution of galaxies on a large scale, through the so-called integrated Sachs-Wolfe effect, and by a number of other observations. Keep in mind that dark energy is a system of a huge number of freely adjustable parameters that facilitate the interpretation of *any* observational result. Nothing describes the situation better than referring to a former utterance of Erwin Schrödinger: "Once the problem is solved by an excuse, there is no need to think about it any longer!" Reflecting on fundamentals has been completely lost within the dark energy business.

Francis Farley, a successful nuclear physicist, yet regarded as an outsider in cosmology, has pointed out[71] that the data of supernovae actually prove that there is no attraction between galaxies; this is exactly the observation of an "empty universe" of which I have spoken. I remember well the first time I heard of this eerie result—it was in 2006 during a lecture by Bruno Leibundgut, a member of a supernova team that had discovered the 'accelerated expansion.' At the time, Leibundgut mentioned that the data initially favored the empty universe interpretation—something nobody could make sense of. Later, researchers abandoned this view because, as Leibundgut stated, the data indicated otherwise.

Chapter12: Einstein's idea is visible long since

At that time, I believed this (I had not realized the link to Einstein's idea of 1911) and stifled my doubts. I allowed the scientific community to convince me that 'dark energy,' which offered many set screws at one's disposal, would describe the data better than the 'empty universe.'

Preferring a complicated model with freely tunable parameters over a simple solution means nothing else but being ignorant of the history and methodology of science. The supernova results are a classic case of the anomaly that Thomas Kuhn described. Whenever a fundamentally wrong but established model, such as the medieval geocentric cosmology, hit data inconsistencies, it were always complicating assumptions that saved the day, rather than questioning the paradigm.[i]

For the same reason, the scientific community did not dare to doubt the validity of conventional cosmology as a whole, though the ad hoc postulate of dark energy clearly had the character of an epicycle much like the ones invented by Ptolemaic astronomy for the description of the Martian orbit. In the current example, there simply was no worked-out alternative.

FOR, HE REASONS POINTEDLY / THAT WHICH MUST NOT, CANNOT BE[72]

In 2012, in an email to Leibundgut, I came back to the question of whether the supernova data might be reconcilable with an empty universe. [ii] He admitted that no model had been worked out concerning that interpretation, though he did say he knew just how such a model should look. From his perspective, he was right, and nobody can blame him for that.

---

[i] History repeats itself with astonishing regularity. Later, scientists seriously debated whether the acceleration of the expansion is going to decrease (arxiv.org/abs/1509.03461).

[ii] This is actually supported by new data published in 2017, www.nature.com/articles/srep35596.

Leibundgut, whose interesting seminars I always enjoyed, is a good example of how upright, smart, and serious scholars ultimately remain prisoners of the system. Certainly, he points out where the data conflicts, and he also raises questions that appear too skeptical to his peers, yet he will not publicly question whether the model that cosmology has worked with for eighty years is based on wrong assumptions. No matter that he is a top scientist, in doing so, he would risk being 'excommunicated' from the field, for it is simply unthinkable in the context of the contemporary model that the expansion of the universe might be an illusion.

Regardless of this, we cannot simply sit down in front of our computers for a long weekend and examine the observational data for compatibility with Einstein–Dicke cosmology. Too much of complex data analysis, involving legions of scientists, has already been done under the assumption of the current model. This brings me to questions of how to properly test theories, which can only be answered in a new culture of open-access data.

> *A knowledge of the historic and philosophical background gives that kind of independence from prejudices of his generation from which most scientists are suffering. This independence created by philosophical insight is – in my opinion – the mark of distinction between a mere artisan or specialist and a real seeker after truth.*[73] *– Albert Einstein*

Chapter 13

The Next Revolution Needs Open Data

Why astronomical precision measurements must go online

While reading this book, you have certainly asked yourself more than once why physicists should not have discovered Einstein's idea long ago if there was something to it. In the last chapter, I pointed out that the dilemmas related to "flatness" and "dark energy" clearly indicate that Einstein's variable speed of light was the right concept for understanding the cosmos. That such a model has a long way to go before receiving recognition will dawn on you in this chapter.

There is a trivial reason for this: The currently established "concordance model" of cosmology also dominates the realm of data evaluation. The analysis of modern Big Science experiments is much too complex to be revised over a short period of time by a couple of mavericks. It would take legions of well-versed scientists at one's disposal to support an alternative point of view with the same kind of technical evidence the standard model provides.

Despite some promising open data approaches, the current situation in science is problematic. As long as a research community that follows a particular paradigm has sole access to the data along with the resources to evaluate it, it is completely illusory that a radical new idea will have any chance of being examined. We are not yet speaking of the insurmountable obstacles in the review process, when papers challenging the mainstream are submitted for publication and evaluated by "experts" of the paradigm. A radical idea would not even succeed in reaching this point.

ANTIQUATED METHODOLOGY

Therefore, it is not only the paradigm of current cosmology that has to change, but also the methodology of science. Perhaps only a technical revolution can bring significant change. Technical revolutions have often caused proper scientific breakthroughs—think of the telescope or electrification. The Internet could lead to a transparent, repeatable, simply better methodology of science. Then, the phenomena raised in the previous chapter would be investigated in a more objective manner; and a number of other observations would instigate the reconsideration Einstein's old idea.

As a general rule, science only works with observations. It is the core of the scientific method that theories have to prove themselves while standing the test with data; theories must also be disprovable. This criterion was emphasized by philosopher Karl Popper, who called it falsifiability. A theory that does not even open the possibility that it fails to agree with observation is just an ideology. Popper's criterion is a wonderful antidote against modern theoretical fantasies, such as string theory, which in decades has not yet succeeded in making a single prediction (its adherents believe the theory to be too beautiful to be wrong).

Falsifiability is therefore necessary, though unfortunately not sufficient, to define good science. In particular, not everything vaguely referring to experiments makes sense. Just take the example of the increasingly more powerful particle accelerators that have evolved to the level of a high-tech sport without requiring a creative idea.[74] History has often shown that notions allegedly supported by observations were inadequate; in some

Chapter 13: The next revolution needs open data

cases they were absolute rubbish. A failure to acknowledge this would be a true denial of reality.

Particularly illuminating in this context are books like *Gravity's Shadow* by Harry Collins, and although on a different topic, *Constructing Quarks* by Andrew Pickering. Both Collins and Pickering demonstrate that experiments and theories are not separate worlds but often develop a symbiosis while describing reality. Concepts such as dark matter and dark energy are, in this sense, of dual use: for observers, they are a tool to organize their data, while theorists consider their work justified by observations and can claim it deserves further investigation. What reality consists of is determined by the consensus of those involved—no more, no less. In this process, it is of secondary importance whether the concepts are adequate representations of nature, and Collins, Pickering, and others have shown that this consensus is essentially a sociological process. If there is an insight (even at meta level) beyond doubt, then it is that.

Ask a practitioner in astrophysics or cosmology and he will also acknowledge the role of sociology, if he is honest and not brainwashed by his own ambitions. Numerous anecdotes full of self-irony refreshingly show that scientists are in no way convinced of the absolute validity of the standard model of cosmology, even if they ritually praise it in the abstracts of their papers. In the back of their minds, at least, there is some residual skepticism.

> *It is easier to destroy a preconceived opinion than an atom.* – Albert Einstein

THE HEAD-IN-THE-SAND PARADIGM

In many scientific communities, a groupthink has spread that blocks individuals from questioning what could endanger the base of their common efforts. I do not mean this is a clear-cut interdiction, but rather a mostly unconscious psychological phenomenon. It is almost impossible to sit and listen to a lecture in a hall with five hundred people and simultaneously question if the subject of the lecture makes sense. You don't want to remove yourself from that community spirit not even mentally; this predilection influences your power of judgment. We do not dare to consider what the majority views unthinkable.

Thomas Kuhn identified very clearly these predominant sociological mechanisms of real academic life. Right up to the present day, except for one or two howls of protest, no one has seriously challenged his theses. Unfortunately, this does not mean that Kuhn's insights have become practice—that is, by means of supervision by scientists outside the community, as to whether the research program met certain methodological standards and so on. Professional institutions, from this perspective, are behaving in a highly unprofessional manner.

A prevailing paradigm, in the Kuhnian sense, prevails for the very reason that observations complying with it are investigated in detail, whereas contradictions, even if they are quite obvious, simply aren't taken seriously. Once they develop to the state of being undeniable, usually a complication, or "extension," of the model is invented, a free parameter designed for no other purpose than to digest the data. In the previous chapter, I outlined the sequence of such complications that have occurred in cosmology, consisting of the notions of expansion of the universe, "space expansion," and finally, "accelerated expansion."

In addition to the problems concerning the Hubble expansion and dark energy, there is another series of observations that contradict the standard model of cosmology. These can also be explained more effectively within the framework based on Einstein's idea. Patience is required here, too—at the moment, the scientific community is unprepared to examine such questions. However, as soon as there is a methodological dawn of hope in the form of publicly accessible data, an array of interesting facts awaits us.

Chapter13: The next revolution needs open data

Incidentally, this is not the place to give an all-encompassing review of existing tests of gravity. Of course, there are impressive observations in astrophysics that are in line with the geometrical formulation of general relativity from which standard cosmology emerged. The fact that some of these observations may go unmentioned here is certainly a good opportunity for malevolent reviewers of this book to bemoan the missing mention of a certain experiment (incidentally, from one's own institute), to allege the author is unaware of it, and to conclude that the arguments made in the book are flawed.

I would like to share a bit of scientific logic with those people: confirmations of the established theory are nice, but they do not reconcile contradictions that have appeared elsewhere. Moreover, they do not say anything about the viability of an alternative that is based on simpler concepts. Now let's list several observations that are a nuisance for the standard model.

GALAXIES THAT ARE FAR TOO SMALL

Many isolated observations that have gone unnoticed (despite proper publication), look quite different from the perspective of Einstein-Dicke cosmology. Data from the *Sloan Digital Sky Survey* (SDSS) galaxy catalog, accessible to everyone is an outstanding example.[75]

A clear indication that there is something wrong with the current model is the size of distant galaxies. The application of the common form of general relativity to the cosmos leads to a muddle of postulates, which were discussed in the last chapter. Accordingly, there would be differing expansions of each space, light, and matter. This is not only completely counter-intuitive, but also unconfirmed by data. In particular, the predictions of the visible size of galaxies, which, according to the model, should increase from a certain distance, was just blatantly wrong. As Martín López-Corredoira has shown in a study[76], the distant galaxies would have to be six times larger than observed in order to meet the expectations.

However, as if it had been anticipated by Kuhn, there is a strong incentive for making use of arbitrary assumptions before one concedes that the model has failed. Accordingly, it has been suggested that galaxies themselves were subject to change in brightness and size (evolution), to the amount, of course, that coincided with the data. Although the contradiction is actually striking, no one makes a great fuss about it. In another study, an additional parameter was introduced, without even bothering to assign a physical explanation to it.[77] Such arguments are difficult to disprove, but from a methodological point of view, they are conjured-up excuses. López-Corredoira commented sarcastically, "evolution is the wild card that solves this kind of problem."

The analysis by López-Corredoiras, however, positively shows a very simple, and in view of variable speed of light, exciting relation: the farther away galaxies are, the smaller they appear—as in everyday life when we look at cars or skyscrapers from a distance. (How the various hypothesized evolutionary effects conspire to create such a simple illusion remains the secret of their advocates.)

Considering the unspoiled evidence, the interpretation that leaps to mind is that there is neither a notable evolution nor an acceleration nor an expansion at all; the angular extension of galaxies visible in the sky simply decreases with distance, as is naturally expected in the Einstein-Dicke model. It is certainly a pity that Dicke never saw this data—he died in 1997, shortly before the flourishing of Internet data.

TWILIGHT SIGNALS

A further, rather repressed topic of cosmology that López-Corredoira also took on is that of quasar anomalies. These virtually point-shaped light sources are almost indistinguishable

Chapter 13: The next revolution needs open data

from stars at first glance, but they show an enormous redshift. They seem to be extra-galactic objects, according to the prevailing interpretation, young, luminous nuclei of galaxies.

There are already noteworthy elements here. While most astrophysical objects are not characteristic for a certain phase during the evolution of the cosmos, it seems there is a preferred epoch for quasars that is absolutely uncommon. López-Corredoira collects a series of serious discrepancies, yet almost resignedly summarizes:

> *There are, however, many papers in which no objections are found in the arguments and they present quite controversial objects, but due to the bad reputation of the topic, the community simply ignores them. This has become a topic on which everybody has an opinion without having read the papers or knowing the details of the problem, because some leading cosmologists have said it is bogus.*

An interpretation of the early times of quasar physics that is considered heretical nowadays would be that the redshifts of quasars are not only cosmological in nature, but also created by the gravity of the object itself.

According to general relativity (the gravitational redshift mentioned in Chapter 8), this would be possible in principle, but there is no established theory to explain it. However, it would solve the serious problem of the quasars' gigantic luminosity, an observation that has generated a sequence of newly postulated mechanisms, each one in line with the latest picture.

If part of the redshift results from gravity instead, the distance derived from the redshift is correspondingly smaller; the luminosity would be less dramatic (a closer object seems brighter to us, even if it is relatively small). In principle, the SDSS offers a huge data record on quasars from which the question could be examined in detail.

Einstein's Lost Key

STRUCTURE FORMATION AND THE LACK OF REASON FOR IT

> *I feel oft en that we are missing some fundamental element in our attempts to understand the large-scale structure of the universe.*[78] – *Margret Geller, American Astrophysicist*

Structure formation in the universe is poorly understood, and therefore, an issue that is heavily underrepresented. What does the term mean? One supposes—and it seems reasonable—that there was no preferred spatial direction and that matter was homogenously distributed in the early times of the universe. That's the way data called cosmic microwave background are interpreted.

On the other hand, matter in the universe is obviously very concentrated. This matter consists of planets that hover and float in the emptiness of interplanetary space, and interstellar space that surrounds solar systems such as ours with a still lower density. But that is not all, stars concentrate in galaxies, and galaxies in galaxy clusters, which are surrounded by huge intergalactic empty zones, so-called voids.

That voids are almost completely depleted of matter is a mystery in itself, but above all, it is astounding that the scheme of smaller structures of matter concentrating to larger ones doesn't stop here. Galaxy clusters form superclusters and super-superclusters, and the statistics show that the size of these structures approaches the scale of the entire visible universe.[i] This is a heavy contradiction of the old model, that over the years had to concede the existence of ever larger structures and has now

---

[i] Francesco Sylos Labini has rendered outstanding services in highlighting these problems (arXiv.org/abs/1103.5974 and 1110.4041.)

Chapter13: The next revolution needs open data

come to the end of the chain of justifications by ignoring the contradictions.

At a galactic level, the first attempt was to solve the problem by postulating dark matter—it was supposed to have concentrated for some unknown reason beforehand and then normal matter followed. This mechanism already sounds contrived, but it wasn't even sufficient to explain the enormous clusters of matter in the cosmos. Then, another free parameter called "bias" was introduced to resolve the discrepancies in the theory of galaxy formation without ever making an effort to find a sensible physical counterpart. Further arbitrary parameters, such as "brightness evolution," did not find their way into the public perception.

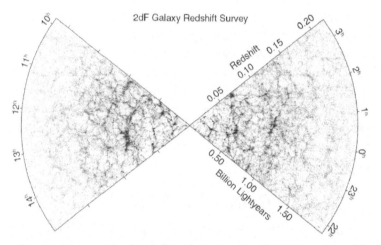

Modern picture of galaxy distribution in the universe, as it was measured in this manner for the first time by Geller und Huchra (1986). Noticeable structures with spatial inhomogeneities appear. Up until now, these have been difficult to explain by the standard model. [Source:2dF Galaxy Redshift Survey Team]

While the unclear understanding of the formation of large structures in the universe has attained at least limited prominence, [79] the fact that there is little comprehension about how the sun and the planets came about has almost been forgotten. Although the idea of gas clouds that contract due to their own gravity seems plausible, this scenario seems far from providing a quantitative and unequivocal comprehension.[80]

Generally speaking, at all size scales, no one really understands how evenly distributed matter can cluster to the concentration that is observed. The anomaly is always a higher degree of organization than expected. The only possible reason is that we are still missing an essential element in our attempts to understand cosmic evolution. At this point, Dicke's model offers an intriguing solution. Due to variable length scales, which, as explained in Chapter 10, were shrinking while the universe aged, matter was practically forced to organize itself.

In other words, the effect of gravitation would have been much larger at the beginning of the universe; if it decreases, according to Mach's principle, more matter would become visible at the horizon. The models of cosmic evolution suffer precisely from the fact that the gravitational constant G has been assumed to be of equal value for all time.[i] Certainly, a gravitational interaction that decreases, as in Dicke's model, has the potential to explain many of these conundrums. In this case, open access to data would provide the prerequisites for resolving such challenges.

G AND THE EARTH

At this stage, the counterargument resurfaces that an increase of the gravitational constant G, as already assumed by Dirac, has been falsified by observations. By using such a quick judgement, it is easily forgotten that many of these presumably pure observations, as portrayed by Kuhn, took place within the

---

[i] I am not advocating that G be freely tunable as with the usual parameters, but be determined by mass distribution.

Chapter 13: The next revolution needs open data

model, and for that reason, scientists failed to take note of the evidence.[i]

It is unclear how these experiments are to be interpreted within the changing scales framework outlined above. However, an equally surprising test for the variability of G would be the tiny increase in the radius of the Earth. If gravitational pull weakens over time, the huge pressure in the Earth's interior that acts against gravity would cause a small expansion. This hypothesis was mentioned for the first time by Pascual Jordan in his visionary work *Gravity and Space*, published in 1955, and since that time, it has in no way been disproved.

The problem is that this idea was, for a period of time, in competition with the continental drift, already predicted by Alfred Wegener in 1912, but not accepted before the direct observation of seafloor spreading around 1960. (Incidentally, this is a glaring example of groupthink within a scientific community.) There is, of course, not the slightest doubt regarding continental drift, which typically amounts to several centimeters each year. However, if the Earth would expand (radially) to a smaller degree, a changing gravitational constant G would cause something like tens of a millimeter annually. The results thereto are as yet unclear.[81] Jordan's extremely interesting geological arguments, as well as the possibility of checking this hypothesis with the grooves on the Moon's surface (they could have been created by expansion), are however valid.

Another possible direct evidence for the trend of G is provided by superconducting gravimeters. These highly-sensitive devices can measure variations of local gravity force down to nm/s^2, which is to say one part in 10 billion! Nevertheless, local gravity is influenced by a number of effects, from the tidal pull of the moon and the sun to the deformation of the earth's crust

---

[i] That applies, for example, to the measurements of the distance to Mars by the Viking probes in 1979.

via high air pressure. Although the noise in the data to date prevents a clear statement, the technology to create a probe of variable G is within reach. In principle, superconducting gravimeter data are available on the Internet.[i]

GRAVITATIONAL WAVES - REALLY DISCOVERED?

Of course, the question also arises to what extent gravitational are evidence against variable speed of light. Before that, however, is interesting, to say the least, to take a look at the history of the gravitational wave discovery – quantitative predictions of their intensity have been continuously adjusted while researchers had remained empty-handed for decades. Obviously, the risk of artifacts increases with the weakness of a signal. The precision is currently said to be 10^{-21} m, one millionth of the proton radius.[82]

However, the circumstances of the discovery itself also raise some doubts. The signal GW150914 presented in the press conference of Feb. 11, 2016 that immediately convinced the world of the existence of the waves, had been measured shortly after commissioning, without safety precautions. To this day, it remains the strongest (!) signal ever measured. The group around Andrew D. Jackson at the Niels Bohr Institute in Copenhagen later found numerous inconsistencies in the data.[83] LIGO had to admit that the central figure in the discovery article had been reworked "for pedagogical purposes."[84] Even after the `confirmation' by another signal in 2017, there are still open questions. Last but not least, the extremely high masses of the corresponding black holes and the numerous false alarms are astonishing. Although the scientific community is largely convinced of the existence of waves, I do not think the discussion is closed.[85]

---

[i] http://ggp.gfz-potsdam.de. Unfortunately access is limited. An attempt in this direction is indicated in my paper arxiv.org/abs/gr-qc/0610028.

Chapter13: The next revolution needs open data

THE METHOD THAT WORKS

Can a new culture of data analysis really take over? In the previous chapter, I addressed a series of observations on dark energy and flatness, which would be better explained by Einstein-Dicke cosmology. However, the problem therein is that the raw data are usually far too complex to be processed by individuals outside of the "community."

The evaluation may be carried out by the leading institutions conscientiously and to the best of their knowledge, however, the trouble lies in the usual way of data evaluation, which is a "vertical" process. This process begins at the first stage of raw data, which, for instance, is delivered by satellites, and continues through various calibrations, filterings, and noise reductions, and eventually compares the data to the theoretical models under consideration. However, the complete evaluation is typically conducted by one single group of scientists, who at the very end proclaim that the universe contains a certain percentage of dark energy. Here is an example:

One of the prestigious projects of cosmology, the cosmic microwave background (CMB), is, unfortunately, a prototype of non-transparency. The data has been mainly evaluated by large teams from the space probes COBE, WMAP, and Planck. Data analysis associated with these missions is impossible to oversee and cannot be repeated without the use of extensive resources; it is also certainly influenced by the framework the paradigm specifies. The public does not have access to the raw data and some valid questions regarding the evaluation have been ignored.[86] The same applies to the 'first picture' of a black hole in the M87 galaxy.

All this does not mean that the results are doctored or distorted, but the fact that the final result is reported in terms of the established model constitutes a selection that makes the entire project (owned by the taxpayer) worthless for alternative approaches. Particularly, fundamental research in particle physics is organized in a bizarre way; the big accelerators haven't released a single dataset useful to the public for decades, while arguing that only experts are qualified to examine the data. It is hoped that scientists migrating from particle physics don't bestow their methods on cosmology.

Fortunately, the situation in astrophysics isn't that messy everywhere. The last decades were a golden age of observation; space-based telescopes in Earth orbits or at Lagrangian points (orbiting the Sun with the Earth as a companion) deliver data of unprecedented quality on all wavelengths. Science, in principle, could flourish, all the more because computers and image processing have caused a revolution of their own in data analysis. However, unfortunately, posting the raw data on the Internet and providing the routines that have been applied is not a matter of course everywhere.

To improve methodology in a generally significant way, a "horizontal" system of data processing would have to be substituted for the "vertical" system described above. Instead of all-in-one, huge collaborations, data analysis must be performed at various stages by specialized, yet independent teams. Of course, satellite navigation and/or signal readout routines from CCD chips are complex issues that require the highest level of skill and precision. But there is no need for the respective groups to communicate their procedures; such a mutual tuning tends to bias the final result. Naturally, during the entire process, it is also absolutely necessary to disclose the source codes of the programs that have been applied.

The raw data (though there are different types of "rawness") must be available online. If it were, there would be a relatively clean intermediate state of data, that is, the luminosity of signals of light at particular positions in the sky at certain wavelengths at definite times. Interpretations would not have been entered so far. Unfortunately, these kinds of data are not automatically useful. At a further stage, it is often necessary to subtract an

Chapter13: The next revolution needs open data

unwanted foreground signal from the data. An independent scientist would want to do his or her own data reduction. At this point, the bulk of assumptions could have already entered the analysis, and it becomes clear once more that there is no naive "realism" and that practically all observations require interpretation.

However, by implementing such a horizontal policy, science would gain a great deal. There are a few successful examples that have shown that such horizontal data evaluation can work. The galaxy (and other objects) catalog *SDSS* or the *Hubble Deep Field* picture of a section of the sky contain top-quality astronomical data that can be accessed by anyone. These datasets have generated an unexpected number of publications. Not by coincidence, I suppose, they have also uncovered the clearest contradictions to the standard model.

WHERE THERE IS MISERY, HOPE IS BREEDING

There is one project about which I am particularly optimistic: *Gaia*. In 2016, this telescope, maybe the most important in the astronautics era, will begin to deliver data. The measuring principle of *Gaia* is over two thousand years old; the ancient astronomer Hipparchus of Nicaea already determined astronomical distances with the parallax, that is, the angular difference under which an object is targeted between two different locations of observation. Because it is so straightforward, the method is less prone to a number of subtle deceptions that astronomical distance measures are notorious for. The most prominent victim of such a deception was Edwin Hubble, whose accomplishments are not belittled by the fact that Hubble underestimated the distances to galaxies by a factor of ten. *Gaia* will establish a new era of precision astronomy[i] worthy of the name. However, the benefit can bear fruit only if the data is publicly available and if it is transparently processed. It seems that *Gaia* will become a prominent example of horizontal data processing and a paragon for future missions.[87]

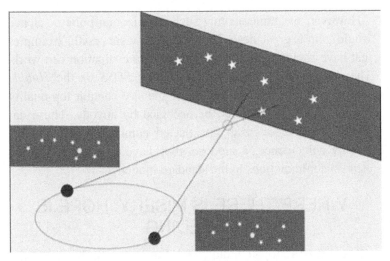

Principle of stellar parallaxes: Due to the annual motion of the Earth around the Sun, the position of a star with respect to the distant background oscillates with a semiannual period (vastly exaggerated here).

Therefore, it seems reasonable that the technical revolution of space telescopes will be properly succeeded by a revolution in data processing based on the horizontal principle. Actually, this would be a return to one of the basic values that defines science: reproducibility. This means that every observation that is considered evidence for a theory may be scrutinized independently by an unlimited number of scholars. Today's communities have allowed the degradation of this imperative prerequisite of science and have transformed what ought to be the source of humankind's knowledge into an expertocracy that has obfuscated

---

[i] If we also apply precision to the notion, it's called *astrometry*.

Chapter13: The next revolution needs open data

true research. Science, however, cannot develop further without providing absolute transparency when presenting its results. Transparency and reproducibility are, however, no longer a distant dream in the Internet age. Sooner or later, science will be compelled to disclose all data and related procedures, making them thoroughly examinable by the public.

Only a methodological revolution of this nature can pave the way for a change in content that is needed for the current paradigm of cosmology. Einstein's idea, properly developed, would eventually shed light on its fundamental mysteries. The concept is simple, intuitive, and in all of its consequences, no less revolutionary than Einstein's previous work. It is time to open this hundred-year-old treasure chest.

Outlook

In Einstein's time, physics flourished. Maybe you happen to agree with me that contemporary physics is in the midst of a crisis. As a taxpayer, you might be interested to know that the billions of dollars spent on real or alleged fundamental research is well invested. Yet, the problem is a more basic one. Our understanding of the laws of nature will decisively contribute to how our civilization will be able to adapt to a cosmos that is rather hostile to life. Therefore, we must question the future development of physics because the destiny of *Homo sapiens* on Earth may be at stake. Every one of us can only hope to provide a modest contribution to the respect that may be granted to us by future civilizations.

Acknowledgment

Again, many thanks go to my family and my friends for their comprehension and for various other forms of support. I feel particularly indebted to my country that has succeeded so far in preserving freedom of the individual, constitutionality, and an open knowledge society. It has permitted me to dedicate a large part of my life to the search for the laws of nature. May it stay like that.

Images

If not specified otherwise, the figures are in the *public domain*, mostly from Wikipedia, or the rights are hold by the author.

Bibliography

Arp, Halton: Seeing Red, Apeiron 1998

Barbour, Julian: The Discovery of Dynamics, Oxford Univ. Press 2001

Barbour, Julian: The End of Time, Oxford Univ. Press 1999

Collins, Harry: Gravity's Shadow, Univ. of Chicago Press 2004

Cornell, James, ed. Bubbles, Voids and Bumps in Time. Cambridge and New York: Cambridge University Press, 1992.

Debever, Robert (Hrsg.), Cartan, Élie, Einstein, Albert: Letters on Absolute Parallelism, 1929–1932, Princeton University Press 1979

Einstein, Albert: The world as I see it (German :Mein Weltbild)

Feyerabend, Paul: Against Method.

Feynman, Richard; Leighton, Robert; Sands, Matthew: Lectures on Physics, Bd. 2, Addison-Wesley 1964.

Feynman, Richard: QED – The Strange Theory of Light and Matter, Princeton Univ. Press 1988

Gutfreund, Hanoch und Renn, Jürgen: The Road to Relativity, Princeton University Press, 2015

Heisenberg, Werner: Der Teil und das Ganze (The part and the whole)

Horgan, John: The end of science, Basic books (1996)

Isaacson, Walter: Albert Einstein – His life and universe. Simon & Schuster Reprint 2008 (Kindle)

Jordan, Pascual: Schwerkraft und Weltall, Vieweg 1955 (German)

Kragh, Helge: Dirac, Cambridge Univ. Press 1990

Kuhn, Thomas: The structure of scientific revolutions

Landau, L.D., Lifschitz, E.M.: Theoretical physics, vol.2.

Lindley, David: The End of Physics, BasicBooks 1993

Lindley, David: Uncertainty.

López Corredoira, Martín: Against the Tide, Universal Publishers 2008

López Corredoira, Martín: The Twilight of the Scientific Age, Brown Walker Press 2013

Mach, Ernst: Die Mechanik in ihrer Entwicklung, historisch-kritisch dargestellt, 1883 (German)

Mackay, Alan: A Dictionary of Scientific Quotations, Adam Hilger 1991

Moszkowski, Alexander: Einstein, Einblicke in seine Gedankenwelt, 1921

Pais, Abraham: Albert Einstein – Subtle is the Lord. Oxford University Press 2005 Kindle (Orig.: 1982;

Parker, Barry M: The Vindication of the Big Bang, Springer 1993

Pickering, Andrew: Constructing Quarks, Univ. Chicago Press 1986

Rosenthal-Schneider, Ilse: Begegnungen mit Einstein, von Laue und Planck, Vieweg 1988 (German)

Sanders, Robert: The Dark Matter Problem, Cambridge Univ. Press 2010

Schrödinger, Erwin: Die Natur und die Griechen, Rowohlt 1956

Schrödinger, Erwin: My view of the world.

Singh, Simon: Big Bang, 2004, Fourth estate

Unzicker, Alexander: Vom Urknall zum Durchknall, Springer 2010 (German); English: Bankrupting Physics (Palgrave Macmillan 2013)

Unzicker, Alexander: The Higgs Fake – How Particle Physicists Fooled the Nobel Committee.

Index

Accelerated expansion 12, 23, 127, 195, 197, 208, 214
Big Bang 4, 22f, 117, 126, 145, 151, 165, 169ff, 175ff, 230
Black body radiation 185f
Black holes 177
Bohr 31ff, 83, 185ff, 189, 194
Cartan 19, 30, 38, 93ff, 131, 158, 229
Constants of nature 32, 41-54, 115, 154f, 162f, 166, 179-184, 188f, 191
Cosmic inflation 23, 37, 49, 90, 117, 130, 181, 205
Cosmic microwave background 16, 66f, 130, 201, 218, 222
Curvature tensor 102, 105
Dark energy 5, 23f, 195, 198, 199f, 208ff, 222
Dark matter 213, 219
Dicke 4, 8f, 15ff, 21ff, 54, 129-149, 163-179, 197f, 202ff, 210, 215f, 220f, 231
Differential geometry 16, 18, 38, 89, 91, 95, 100
Dirac 4, 8, 9, 15, 19ff, 32, 35, 44, 48, 83, 129, 146, 149-183, 189, 220, 229, 231
Eddington 15, 18, 89, 109, 110f, 154, 156, 161f
Equivalence principle 59f, 64ff, 68, 71f, 76, 79f, 84
Fermat 73ff, 82
Flatness 23, 130, 155f, 203, 205, 207, 211, 222
Freundlich 109
Gaia 224
Galaxies 14f, 22, 116, 151, 168, 182, 195, 197f, 200ff, 207f, 215ff, 225
General relativity 11, 13f, 16ff, 21, 35, 59f, 66ff, 77ff, 89f, 98, 102ff, 108, 111-115, 122, 125ff, 130, 132f, 138, 140ff, 157, 162, 170, 173, 199, 215ff
Gravitational constant 13, 15, 19ff, 26, 46, 49ff, 65f, 77f, 85, 113-125, 130, 134ff, 146f, 158ff, 165, 172, 179ff, 220f
Gravitational potential 15, 77f, 118, 122, 137, 146, 169, 204
Grossmann 16f, 38, 101f, 106ff
History of science 8, 15, 30f, 40, 64, 90, 149, 180f, 188f
Hubble 15, 19, 20ff, 115, 127, 15ff, 162, 165ff, 177f, 195ff, 214, 224f
Kepler constant 113
Kinetic energy 23, 32, 52f, 58, 145, 181, 185, 203f
Kuhn 26, 185, 188f, 193, 206ff, 214ff, 220, 229
Large Number Hypothesis 4, 149f, 157ff, 163, 174, 182
Leibniz 93, 95
Light deflection 77ff, 84f, 107, 110f, 130, 166
Lord Kelvin 110f
Lorentz 59, 80, 86, 188
Mach 3f, 8f, 14ff, 19f, 25f, 32, 35, 54f, 62-69, 79, 82, 85, 113-126, 13ff, 137, 156f, 165, 174f, 179, 189, 202, 220, 230
Mach's principle 65, 124, 156

Maxwell's equations 54, 167
Methodology of science 67, 201, 209, 212, 223
Newton 4, 13, 18, 24, 31ff, 46, 51ff, 62-69, 74, 93ff, 109-118, 124, 136, 143, 146ff, 151, 179-192
Perihelion shift 109, 138, 141f, 144f
Planck 31, 39, 42, 45, 48f, 181ff, 185f, 189, 193, 201, 222, 230f
Potential energy 117, 145, 203f
Quasars 216f
Radar echo delay 138, 140f
Redshift 22, 138f, 151, 167, 168, 172, 177, 197ff, 217
Riemann 82, 100ff, 105, 106
Schrödinger 8f, 35, 83, 90, 117-126, 129, 133ff, 149, 179, 18ff, 208, 230
Schwarzschild 105f
Sciama 8f, 114, 134ff, 146f, 172
SDSS 215, 217, 224
Sociology of science 37, 116, 200, 213
Special relativity 14, 62, 74ff, 145, 149, 188, 204
Standard model 137, 151, 156f, 211ff, 219, 224
Supernovae 195ff, 208f
Time dilation 58, 188
Voids 218

Endnotes

[1] *Annalen der Physik* 35 (1911), p. 905, not to be confused with *Annals of Physics*.
[2] R. Dicke, Reviews of Modern Physics 29 (1957), pp. 363–376. See A. Unzicker, vixra.org/abs/1510.0082.
[3] A. Einstein, Mein Weltbild, p. 34.
[4] This is a matter of debate among historians, see. T. Sauer, Arch. Hist. Ex. Sci. 59 (2005), 577–590.
[5] Shown by Broekaert, FoP 38 (2008), p. 409–435, see footnote 70.
[6] P.A.M. Dirac, Proc. Roy. Soc., London, 165 (1938), p. 199 ff.
[7] I. Rosenthal-Schneider, Reality and Scientific Truth: Discussions With Einstein, Von Laue, and Planck, Wayne State Univ. Press (1981), p. 27.
[8] Since 2015 available online at einsteinpapers.press.princeton.edu/vol3-trans/393.
[9] M. J. Duff, L. B. Okun, G. Veneziano, arxiv.org/abs/physics/0110060.

10. Moszkowski, loc.1719.
11. Mathematische Annalen 102 (1930), pp. 685-697.
12. Élie Cartan and Albert Einstein: Letters on Absolute Parallelism, 1929-1932, Princeton Legacy Library, 2015.
13. Rosenthal-Schneider, p.41; The Max Planck quote shows how fundamental her question regarding the constants of nature was.
14. In his book QED: The strange theory of light and matter (Penguin, 1985).
15. Rosenthal-Schneider, p. 29.
16. Jahrbuch der Elektrizität und Elektronik 4 (1907), p. 433.
17. In a beautiful story in a ship's hull comp. e.g. Singh (2005), p.54.
18. A. Einstein, The World as I see it.
19. *Annalen der Physik* 49 (1916), p. 771.
20. Regarding the problem of "Newton's Bucket", a very interesting article by Donald Lynden-Bell and Jonathan Katz (arXiv:astro-ph/9509158).
21. Annalen der Physik 35 (1911), p. 906.
22. https://en.wikipedia.org/wiki/Second
23. Ellis, arxiv.org/abs/astro-ph/0703751 and the comment arxiv.org/abs/07082927.
24. Pais, loc. 5999.
25. https://link.springer.com/chapter/10.1007/978-3-540-87777-6_2
26. https://einsteinpapers.press.princeton.edu/vol7-trans/156
27. W. Isaacson, (kindle ed.) loc. 3173. Einstein was thus concerned with atomic theory and radioactivity. The congress took place from 30 October until 4 November (Pais, kindle ed loc. 5926).
28. cf. e.g. the sporadic mention in correspondence, http://einsteinpapers.press.princeton.edu/vol5-doc
29. Vierteljahresschrift f. gerichtl. Med. und Sanitätswesen 44 (1912), p. 27-40, https://einsteinpapers.press.princeton.edu/vol4-doc/198.
30. We refer here to the distant parallelism theories of Einstein and Cartan around 1930.
31. A. Einstein, Mein Weltbild (German), Ullstein 2005.
32. See., e.g. Landau-Lifshitz II, § 81ff.
33. www.deutschlandfunk.de/allgemeine-relativitaetstheorie-der-wertvollste-fund-meines.740.de.html, author's translation.
34. F. Winterberg: *Zeitschrift für Naturforsch. A.* Bd. 59, 2004, p. 715.
35. P. Coles, arxiv.org/abs/astro-ph/0102462.
36. Comp. Parks, Phys. Rev. Lett. 105 (2010), p. 110801; arXiv:1008.3203, or the overview arxiv.org/abs/gr-qc/0702009.

[37] D. Sciama, MNRAS, Vol. 113, p.34
[38] See Chapter 10 in *Bankrupting Physics*; Penrose Chap. 27.6.
[39] http://www.heise.de/tp/artikel/44/44441/1.html (German) or http://blog.alexander-unzicker.com/?p=119.
[40] Annalen der Physik 382 (1925), pp. 325–336.
[41] http://onlinelibrary.wiley.com/doi/10.1002/phbl.19600161104/pdf
[42] A. Einstein, Annalen der Physik 55 (1918), p. 241.(not to be confused with Annals of Physics).
[43] A. E. Chubykalo, arXiv:hep-th/9510051, A. Unzicker, arxiv.org/abs/gr-qc/0308087.
[44] R. Dicke, Review of Modern Physics 29 (1957), p. 363-376.
[45] R. Dicke, Review of Modern Physics 29 (1957), p. 365.
[46] D. Sciama, MNRAS, vol. 113, p.34.
[47] C. F. Will, relativity.livingreviews.org/Articles/lrr-2014-4/
[48] The Pound-Rebka Experiment, for the foundation of which the Nobel Prize of 1961 was awarded.
[49] J.L. Snider, Physical Review Letters, vol. 28 (1972), pp. 853-856.
[50] Yilmaz, Huseyin, Physical Review, vol. 111 (1958), pp. 1417-1426.
[51] H. Dehnen et al., Ann. Phys. 461(1960), pp.370-406; K. Krogh, ArXiv.org/abs/astro-ph/9910325; M. Arminjon, ArXiv.org/abs/gr-qc/0409092; H. E. Puthoff , ArXiv.org/abs/9909037.
[52] J. Broakaert, arxiv.org/abs/gr-qc/0405015
[53] See also A.Unzicker, Ann. Phys. (Berlin) 18 (1), 57-70 (2009), https://arxiv.org/abs/0708.3518.
[54] A. Unzicker and J. Preuss, arxiv.org/abs/1503.06763
[55] Comp. Kragh, p. 187.
[56] P.A.M. Dirac, *Proc. R. Soc. Lond.* A *1938* 165 199-208.
[57] see *Bankrupting physics*, appendix.
[58] http://vixra.org/abs/1301.0110.
[59] No reference in the biographies of G. Farmelo and Helge Kragh.
[60] English: A. Unzicker and T.Case, arxiv.org/abs/physics/0503046.
[61] J.-P. Uzan, Rev. Mod. Phys. 75 (2003), p. 403; arXiv.org/abs/hep-ph/0205340.
[62] M. Mamone Capria, Physics before and after Einstein (IOS Press, 2005), p.156.
[63] P.A.M. Dirac, *Proc. R. Soc. Lond.* A165 (1938) 205.
[64] Dirac, Nature, Volume 192, Issue 4801, pp. 441 (1961)
[65] Arxiv.org/abs/1402.0132.
[66] http://www.astro.ucla.edu/~wright/sne_cosmology.html.

[67] http://arxiv.org/pdf/astro-ph/9805201.pdf
[68] E.g., Richard Smith, http://www.ncbi.nlm.nih.gov/pmc/articles/PMC1420798/.
[69] Müller, Roland, Fritz Zwicky, Baeschlin 1986, p. 427 (German).
[70] http://www.archivefreedom.org.
[71] Farley, arXiv.org/abs/0901.3854
[72] A quote by the German poet Christian Morgenstern.
[73] Einstein to Thornton, 1944, http://plato.stanford.edu/entries/einstein-philscience/
[74] See my book *The Higgs Fake*.
[75] Try it yourself. With very modest means, one can determine an anomaly regarding the size and brightness of galaxies: arxiv.org/abs/1011.4956.
[76] M. López Corredoira, Int. J. Mod.Phys. D 19, P. 245 – 291, arXiv.org/abs/1002.0525; arxiv.org/pdf/1501.01487v1.pdf.
[77] M. Blanton et al., Astroph. J. 592 (2003), P. 819 – 838, arXiv.org/abs/astro-ph/0210215.
[78] Cornell, chap. 3, p.73.
[79] E.g. P.Kroupa, arXiv.org/abs/1204.2546.
[80] See, e.g., van Flandern: *Dark Matter, Missing Planets and New Comets*, North Atlantic Books, 1999 (2nd ed.)
[81] Wu, X. et.al. Geophysical Research Letters, Volume 38 (2011)13, CiteID L13304 Zelensky, Nikita P. et. al. Journal of Geodesy, Volume 88, (2014), 517–537.
[82] www.ligo.org/science/Publication-S5StochDirectional/
[83] arxiv.org/abs/1706.04191; arxiv.org/abs/1802.00340; arxiv.org/abs/1903.02401; arxiv.org/abs/1802.10027.
[84] New Scientist, 1.11.2018 "Grave doubts over LIGO's discovery of gravitational waves".
[85] Further links: medium.com/@aunzicker/five-years-of-gravitational-waves-a-chronicle-of-strange-coincidences-7d22be19319d
[86] Robitaille, a radiological analysis, www.ptep-online.com/index_files/2009/PP-19-03.PDF and www.ptep-online.com/index_files/2007/PP-08-01.PDF.
[87] https://gea.esac.esa.int/archive/.

Made in United States
North Haven, CT
09 September 2024